Lady Soldiers

Lady Soldiers

Lyn Kane Dale

Dear Stephi
I hope you enjoy these stories.
It is lovely to be in touch
again.
Love —
Lyn xx

Erin Productions

Cover design: Silvia Rosenstreich
Book design & production: Silvia Rosenstreich
Editing: Sandra Wilson
Author photograph: Heather Simms
Typeset in Minion Pro by Silvia Rosenstreich
Printed in Western Australia by Scott Print

National Library of Australia Cataloguing-in-Publication entry:

Author:	Dale, Lyn Kane, 1945- author.
Title:	Lady Soldiers / Lyn Kane Dale.
ISBN:	9780992545604 (paperback)
Notes:	Includes bibliographical references.
Subjects:	Australia. Army--Women--Biography.
	Australia. Army. Women's Royal Australian Army Corps--History.
	Australia. Army. Royal Australian Army Nursing Corps--History.
	Women soldiers--Australia--Biography.
	Vietnam War, 1961-1975--Veterans--Australia.
	Post-traumatic stress disorder--Australia.
	Women veterans--Family relationships--Australia.
	Veterans' families--Australia.
Dewey Number:	355.3480994

Acknowledgements
The publishers would like to thank the Department of Defence for permission to reproduce the 1963 and 1969 WRAAC recruiting brochures.

Dedicated to:

Cherryl Burrows Payne

Helen McLeod Meyer

and all other women

who have served in the Australian Army

In memory of

WX3150 Pte John Murray Kane (1909-1988)

2/16 Battalion A.I.F.

Contents

Acknowledgements

I would like to express my thanks to everyone who has assisted me in the production and launch of this book:

✻ My daughter and son-in-law, Karen and Chris Carmichael for their much appreciated help, love and encouragement. My brother, Murray Kane and cousins, Yvonne Kane James and Wendy Naughton Low, for their interest in the *Lady Soldiers* projects.

✻ Kath Balfour for her valued friendship and for believing in me and this project.

✻ Silvia Rosenstreich for her practical advice and help and for editing, proofreading and formatting the manuscript, for her beautiful artwork, and above all for her friendship.

✻ Alec McHoul who has helped in myriad ways from scanning, repairing and re-sizing photographs to keeping my computer running. Many thanks, Alec.

✻ Sandra Wilson for her friendship, help, encouragement and suggestions over countless cups of coffee, and for the hours she spent reading and correcting the manuscript.

✻ Department of Veterans' Affairs *Saluting Their Service* for the grant which has helped defray some of the costs of producing this book.

✻ Partners of Veterans Association of Australia - South Australia Branch Inc. for providing a donation towards the printing costs of this book and for bringing public attention to the lives of those women who married Veterans. A special thank you to Helen Adamson.

✻ Claudia Cream, OAM, for her generous donation towards printing costs.

✻ Major Chris Roe, Manager of the Army Museum of South Australia, Warrant Officer Jim Love, Curator, and the staff and volunteers for their valuable support, assistance and for hosting the launch of *Lady Soldiers* in Adelaide. Special thanks also to Jenny Anderson, Heather Simms and Trish Ranford for their valuable support and assistance.

✻ Major Barry Willoughby, former Manager of the Army Museum of South Australia, for helping to pave the way for the *Lady Soldiers* projects and for his continuing support throughout the journey.

✻ Veterans SA (Department of Treasury and Finance, Government of South Australia) for their valued assistance.

✻ Adelaide Legacy and Geoff Banks, Development Manager, for their belief in the value of this project and their much appreciated support.

✻ The City of Cockburn (WA) and Christine Elaine, Curator of the Azelia Ley Museum, for hosting the launch of *Lady Soldiers* in Perth.

✻ May Tomich, President, and Jean Pearce, Treasurer, of the WRAAC Association of WA who have administered the funds for the grant.

✻ Helen Flavell for her help, encouragement and suggestions for writing the grant application.

✻ Steve Kinnane for his friendship, advice and enthusiasm for this project.

✻ Denise Hudson of WRAAC Reunited, Pauline van Buren, members of WRAAC Facebook, Betty Hall Layt and Gai Hutchison of the WRAAC Association of NSW, all of whom helped in publicising the book and helping to find participants for the project.

✻ My Murdoch colleagues: Kai-Ti Kao, Mick Broderick, Felicity Newman, Mark Busani, Karen Olkowski, John McMullan, Michael Done and Simon Order for their encouragement and friendship.

✻ Amanda Third for her friendship over many years, for launching the book in Adelaide, and for her support and interest in the Lady Soldiers' stories and other projects that we have worked on together.

✻ Yvonne Henderson for launching *Lady Soldiers* in Perth. Many thanks, Yvonne.

✻ Rob Morrison from Scott Print (Perth) for producing this beautiful book. Thank you, Rob!

✻ Leon and staff at Café Lumos in White Gum Valley (WA). Thank you all for the great coffee and friendship, and for not charging me rent for the many hours I have spent occupying a table in the café while working on this book!

✻ To Cherryl Burrows Payne. Thank you for everything – your much valued friendship over the past fifty years, your enthusiasm for the film and book projects, for typing transcripts, reading and critiquing the manuscript, and for accompanying me on 'road shows' to far-flung places in WA and beyond. And you thought retirement was going to be peaceful!

✻ To Helen McLeod Meyer. What would I have done without your help in negotiating the way through the many miles of bureaucratic red tape over the duration of productions of the film and book? Thank you so much for attending meetings, cajoling, sourcing help and funding for our *Lady Soldiers* projects, for your organisational skills and for your advice, hospitality and friendship.

And, last but not least.

✻ To all of the Lady Soldiers – thank you for your enthusiasm and generosity in sharing your stories and for trusting me with them. My life has been greatly enriched by the friendship I have experienced while working on this project. Many of you I have never met in person but, because of our shared experiences, you have become valued friends. Esprit de corps.

Australian Government
Department of Veterans' Affairs

opportunity
of a
lifetime
in the
WRAAC
WOMEN'S ROYAL AUSTRALIAN ARMY CORPS.
WRAAC

ROYAL AUSTRALIAN ARMY NURSING CORPS (RAANC)

Foreword

Professor Sandra Wilson

Murdoch University

This book contains the stories of fifty-one women who joined the Women's Royal Australian Army Corps or the Royal Australian Army Nursing Corps between 1953 and 1976, and went on to serve anything between two and twenty-four years. Most of them enlisted in the 1960s and served between three and six years. This was a unique period in Australian history, when women had their own branch of the Army. Women had been serving in the Australian armed forces since 1899, but originally they were restricted to nursing duties. During the Second World War, they began to play a wider range of roles. Women's branches of the military were disbanded at the end of the war, however, and it was not until 1951 that permanent structures in which women could serve in the Army were founded. The Women's Royal Australian Army Corps continued until 1985, though integration into the Army at large began in the 1970s. The Nursing Corps still continues.

The world represented by most of the stories in this book was different from today's world. Opportunities for young women who wanted to do something with their lives were often restricted, and the Army provided one avenue for them. The women who tell their stories here joined up for a variety of reasons. They do not usually say their aim was to serve their country. Rather, they were attracted by three other things. First was the uniform. This is not surprising if you look closely at the publicity images used in the Army's recruiting literature - everyone looks like a model! Young women did not generally have extensive wardrobes and the uniforms they saw were smart, custom-made and of good quality. They wanted to wear them. Second, the Army promised interesting jobs compared with what might be on offer for young women elsewhere. The women whose stories appear in this book drove all sorts of vehicles, became experts on drilling, and did signals work. Even if they did more traditional jobs like nursing, clerical work or mess duties, they were away from home, wearing a uniform with pride and taking responsibility in ways they could not have done at home. The third reason they give for joining up was a desire for travel, adventure and excitement. One former soldier, Sheila Cusack, says the Army was 'an adventure for young women'. WRAAC members did their training in Victoria in the 1950s, and in Sydney from the early 1960s; the trip to basic training was itself a considerable journey for many. One contributor to this book ended up as a nurse in Vietnam, another did signals work in Singapore, a third went to Britain on an Army exchange. Others went to live and work in cities and country towns all over Australia.

It was a different world, too, in that most of the women featured here had to leave the Army when they married. It is an indication of how much the Army meant to them that almost all of them express regret about this. They wanted to marry, but they did not want to give up the independence, the friendship, the confidence and the fun associated with serving in the Army. On the other hand, they often say that military training meant they were able to move on to much better jobs in civilian life than would otherwise have been available to them. Their Army record gave them credibility; employers valued their skills, discipline and work ethic; and the women themselves had the confidence to apply for jobs they would never have aspired to before. As Julie Elverd says, after her Army life, 'I felt I could do anything'. Almost all the women represented here say the Army gave them skills they have used and valued for the rest of their lives, from ironing to time-management to truck-driving. The Army, they believe, gave them confidence, self-esteem, discipline and a sense of belonging. Christina Morgan comments that military service gave her 'the ability to go anywhere and feel completely at ease'. Even if they did leave when they married, being in the Army made a difference to their whole lives.

Not everything about post-Army life was positive. The shadow of the Vietnam War looms over the accounts of the later lives of many women whose stories appear here. Women who served in the Army often married fellow soldiers, and in the 1960s and early 1970s, that might mean their husbands went on to serve in Vietnam. One woman who contributed to this book lost her husband in Vietnam, and many struggled to help husbands returning from the war to adjust to civilian life again, not always successfully, and to weather crises in physical and mental health. Not all marriages survived the strain.

Few people outside or even inside the military know much about the women's corps in the Army, so this book provides an especially valuable record. The personal histories recounted here offer a rich and important historical source that can describe things in a way that other documents cannot. When people tell their stories in their own way, reflecting on their own experiences, we learn things that hardly ever appear in more formal documents: not only how it was, but how it *felt* to be there and to do those things. We hear about the private things and the funny things, which are often left out of official records. We get a sense of how exciting it was for a teenage girl from a country town, or far-flung Western Australia, to go to Sydney, live with other girls, stride out in a smart uniform, and do an important job. There is also a real sense of how much fun it all was, at least most of the time. Not everything was fun, by any means, and future years held a lot of grief for some of these women. While they were in the Army at home, however, the 'lady soldiers' had a great capacity to enjoy themselves.

Lyn Dale is to be congratulated on this fine book. It is a remarkable feat, and has required a great deal of time and a lot of different skills. Lyn has brought together the stories of fifty-one women from all over Australia and has told their story beautifully. She has been tenacious, diplomatic and highly committed to the project. The result is a wonderful record of what it was like to be in the Women's Royal Australian Army Corps and the Royal Australian Army Nursing Corps in the 1950s, 1960s, 1970s and beyond, and of what happened to the former soldiers once they left military service.

YOUR FLIGHT TICKET
AB 241610
You're in good hands–all the way
PTE. L. KANE
–when you fly ANSETT-ANA
SYSTEM OF THE GOLDEN JETS

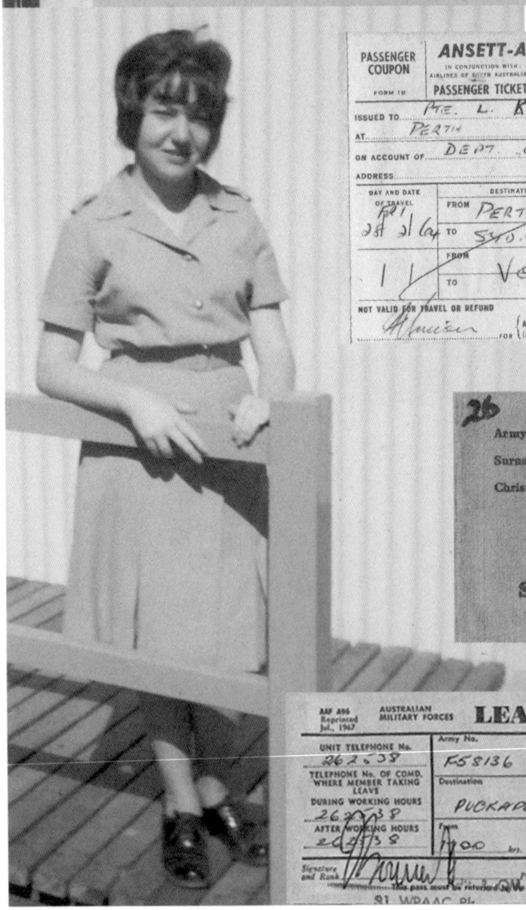

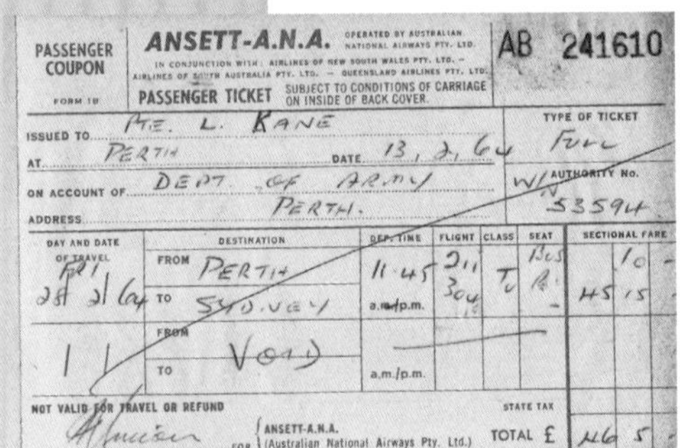
PASSENGER COUPON
ANSETT-A.N.A. OPERATED BY AUSTRALIAN NATIONAL AIRWAYS PTY. LTD.
IN CONJUNCTION WITH: AIRLINES OF NEW SOUTH WALES PTY. LTD. – AIRLINES OF SOUTH AUSTRALIA PTY. LTD. – QUEENSLAND AIRLINES PTY. LTD.
PASSENGER TICKET SUBJECT TO CONDITIONS OF CARRIAGE ON INSIDE OF BACK COVER.
AB 241610
ISSUED TO PTE. L. KANE
TYPE OF TICKET FULL
AT PERTH DATE 13, 2, 64
ON ACCOUNT OF DEPT. OF ARMY
AUTHORITY No. W/N 53594
ADDRESS PERTH.
DAY AND DATE OF TRAVEL
DESTINATION
DEP. TIME
FLIGHT
CLASS
SEAT
SECTIONAL FARE
FRI 28 2 64
FROM PERTH
TO SYDNEY
11.45 a.m./p.m.
211 304
TV
BUS
45 15
10
FROM
TO
VOID
a.m./p.m.
NOT VALID FOR TRAVEL OR REFUND
STATE TAX
FOR ANSETT-A.N.A. (Australian National Airways Pty. Ltd.)
TOTAL £ 46 5

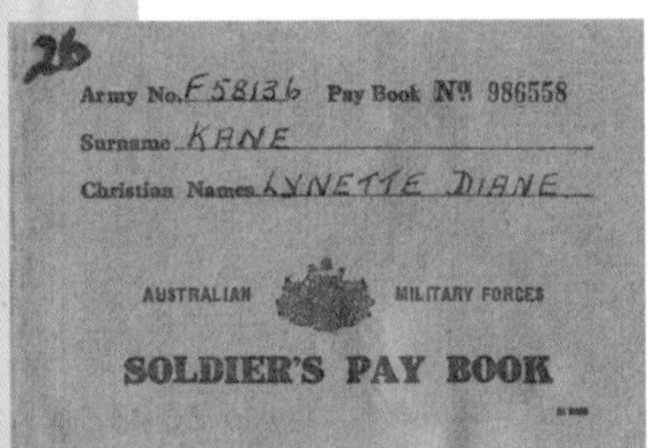
26
Army No. F58136 Pay Book Nº 986558
Surname KANE
Christian Names LYNETTE DIANE
AUSTRALIAN MILITARY FORCES
SOLDIER'S PAY BOOK

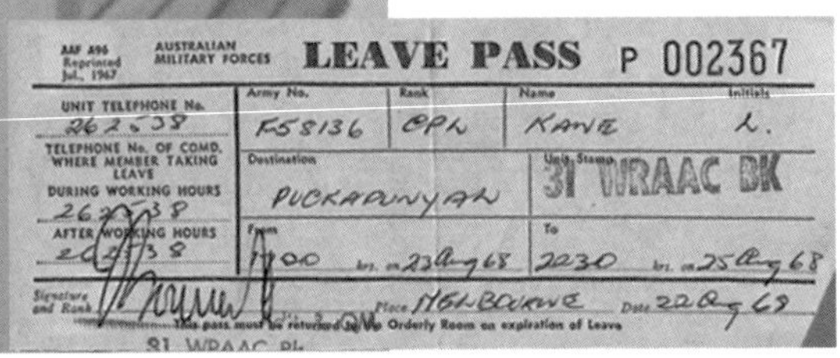
AAF A96 Reprinted Jul., 1967
AUSTRALIAN MILITARY FORCES
LEAVE PASS P 002367
UNIT TELEPHONE No. 262538
Army No. F58136
Rank CPL
Name KANE
Initials L.
TELEPHONE No. OF COMD. WHERE MEMBER TAKING LEAVE
DURING WORKING HOURS 262538
AFTER WORKING HOURS 262538
Destination PUCKAPUNYAL
Unit Stamp 31 WRAAC BK
From 1700 hrs. on 23 Aug 68
To 2230 hrs. on 25 Aug 68
Signature and Rank
Place MELBOURNE
Date 22 Aug 68
This pass must be returned to the Orderly Room on expiration of Leave
31 WRAAC PL

Introduction

A CHANCE TO MAKE THE MOST OF YOUR LIFE

Have you ever paused to think just what kind of life the WRAAC offers you today? Opportunities that only the very fortunate can find in civilian employment.

* thorough training in a wide variety of skills and duties.
* interesting and important work.
* the chance to travel and see Australia.
* wonderful companionship with girls and men of your own age.
* excellent pay and conditions.
* first class opportunities for promotion.

In short, a life that gives you the chance to make the most of your youth and energy.

In February 2011, my friend Cherryl Burrows Payne and I joined some 450 ex-Servicewomen who had come to the former WRAAC School at Georges Heights (Mosman, Sydney) from all over Australia to attend the 60th anniversary celebration of the formation of the Women's Royal Australian Army Corps (WRAAC). Many of the women had not been back there since they completed their recruit training. Cherryl and I had both enlisted in 1964, in different intakes, so it had been many years since we had last seen WRAAC School although the area still looked very familiar to us. The Army handed the land over to the local Mosman council some years ago and it is now a community centre, but a number of the barracks buildings have been retained. The parade ground is now a grassed area instead of being covered in bitumen as we remembered it.

Cherryl tells how it felt to be there.

I gaze around at all the faces waiting for the Company Sergeant Major to call us to attention. We've all come from near and far. 'ATTENTION!' Oh dear, that wasn't good! Much giggling at this stage! 'Again.' Not much better! More giggling!!! 'Again.' – not bad!! This time good. All in unison, here we are marching as one to our regimental tune, *Soldiers of the Queen*. Each of us feeling quite emotional, we're seventeen years old marching up Best Avenue to the parade ground in front of the Commanding Officer and other VIPs. I'm wishing my family could be here to share this special moment with me. I'm wearing my dark green uniform with the gold badges shining and my spit polished shoes. I'm thinking of the reasons that brought me here, wanting more from my life, wanting to travel, not wanting to 'settle down' yet.

But it's not 1964 it's 2011, and today we're here to celebrate the 60th Anniversary of the formation of the WRAAC. The Chief of the Army, Lt. General Gillespie, and the Governor of NSW, Professor Marie Bashir, plus other important people are all in attendance as well as the 'girls' who are unable to march today and are sitting, watching from the sidelines. As we march off the parade ground a male ex-soldier stands and salutes us as we pass by. What a lovely compliment to us all. The Governor unveils the memorial plaque and our flag is lowered for the last time at WRAAC School. It's an awesome feeling just being here.

F58160 Private Cherryl Burrows

Coming together in Sydney for the 60th Anniversary celebrations revived old friendships and memories of service, and new acquaintances were made because of similar experiences. I attended this event hoping to meet up with old friends and was a little disappointed that I only met three I knew. In the back of my mind for many years had been the thought that our WRAAC and RAANC history should be preserved, but as so many women had changed their names when they married it seemed to be an almost insurmountable task to track them down. The 60th celebration gave me that opportunity. I handed out flyers asking women to get in touch with me to talk about their Army experiences. My plan was to make a documentary film to give these women a chance to be heard and tell their unique stories of an era when most young women chose very traditional careers out of the few that were available to them. After my return from Sydney a number of women contacted me. Some just wanted to reminisce and were not interested in having their stories recorded, but others were very keen for their histories to be preserved. Eventually I chose the stories of sixteen women who served between the 1950s and the 1990s. Some of them served during the transition period in the 1980s when the WRAAC was being absorbed into the mainstream Army and was no longer a separate corps. All but three of these women were teenagers when they enlisted.

The majority of the women I interviewed said they had not told their families much about their Army service. Many said they had never spoken of it at all and were grateful to at last have the chance to talk about that period of their lives. They told of the excitement at leaving home for the first time and the great adventure for many of them of travelling to Sydney for basic training. Once there, they had to deal with homesickness and the difficulties of adjusting to Army life. They told of the training they received, learning to live in barracks with complete strangers, making friendships of which many have endured over the years, doing jobs they loved, and being given responsibility at a very young age. A nurse told of working in a military hospital in Vietnam nursing wounded soldiers during the Vietnam War. Another young woman worked as a Secret Cipher in Signals in Singapore in the 1970s relaying messages to and from the war in Vietnam. Others were scattered all around Australia and they told of interesting, funny and sometimes sad times. They also told of romances and marriages and sadness at having to leave the Army when they wanted to marry. Some told of life after the Army with many being Army wives and some going on overseas postings with husbands and children. One young woman's husband died of malaria in Vietnam without ever seeing his baby daughter who was born two days after he left Australia. His twenty-four year old widow was left to bring up their child on her own. A number of others spoke of problems associated with being married to Vietnam veterans and in a number of cases, marriage difficulties and breakdowns.

The common thread running through their stories is the camaraderie amongst the women and the fact that they had to stand on their own two feet at a very young age. After their discharge – in most cases to get married - they put this short but intense period of their lives out of their minds and concentrated on their husbands and families. Some joined ex-Service associations but many lost contact with Army friends and memories of Army days faded or stayed at the back of their minds. In later years many, like me, traced old friends through WRAAC Reunited and were able to reconnect with them.

I applied for a grant from the Department of Veterans' Affairs programme, *Saluting Their Service*, to make a documentary film and was lucky and privileged to receive one. I still had a very limited budget to work with; but friends, work colleagues and family helped with many tasks ranging from scanning and repairing old photographs, to creating spreadsheets and sourcing band music of the WRAAC marching tune, *Soldiers of the Queen*. By this time I had been retired for more than a year. However, my former boss in the School of Media Communication & Culture at Murdoch University, where I had worked for twenty-one years, very kindly gave me permission to use the recording studios in the School so I could record the women's stories. I was also extremely fortunate to have the expertise and enthusiasm of Michael Done, Audio and Web Officer, who did all the recordings and editing.

I contacted cinematographer, Samantha (Sam) Bergersen, with whom I'd worked some years previously. Sam and I met in 2004 when she was a film student looking for a woman's story to make into a film to complete her Media Studies (Film) degree. I gave her a copy of the biography I had written of the life of my grandmother, Maggie Kelly Kane, who emigrated to Australia in 1900 on a 'bride ship'. Sam read it overnight and was back in my office the next morning to say 'I love the story. Let's make a film together'. We got to work, made a forty-five minute documentary film and called it *Maggie's Journey*. Sam and I are a good team who work well together and we are both passionate about telling the stories of ordinary people, especially women. Sam was very enthusiastic about making another documentary – this time about female soldiers, and was keen to get started on the project. We used the recorded voices of the women overlaid with visual context comprising photographs of them taken during their Army service plus Army recruiting posters and brochures, newspaper clippings, letters, postcards and other memorabilia. Nine months later, in April 2012, eleven of the sixteen women whose stories were recorded travelled from all over Australia to attend the premiere of 'their' film, *Lady Soldiers.* It was launched at the Gratwick Club at Campbell Barracks in Swanbourne (WA) by Dr Sandra Wilson, Professor of History at Murdoch University, with an introduction by Major Barry Willoughby, Manager of the Army Museum of South Australia. As one of the Lady Soldiers commented: 'Who'd have thought we would be film stars at our age!'

After the film premiere there remained a great deal of interest in the *Lady Soldiers.* I did not advertise the film other than having it reviewed in the *Army Newspaper,* which resulted in fifty copies of the DVD being sold within 24 hours. The rest of the sales were through word of mouth. Women still kept contacting me and asking if I wanted to hear their stories. I had a lot of material left over from the recordings which had been done for the film so I decided to write a book. Postings on WRAAC Reunited and WRAAC Facebook websites caused many more women to contact me and fifty-one of their stories are told here.

The majority of women whose stories are told in the book served in the 1960s, although there are a small number who served in the 1950s and from the 1970s onwards. These women tell stories of Army life when women had their separate corps – WRAAC and RAANC. They were trained by women in an all-female training establishment, and after their training was over and they were posted to their various units all over Australia they still retained their corps badges and uniforms. Although they were often surrounded by men in work and social situations, the women were housed separately from male soldiers. This probably saved many of them from the type of sexual harassment that some women who enlisted in later years experienced.

To attract young women into enlisting, the recruiting brochures from the 1960s (two of which are reproduced in this book) quite clearly indicate that there would many opportunities to meet and socialise with young men. An Army publicity photograph of Desiree Harms taken in Melbourne in the mid-1960s shows her walking past a sailor who is sitting on a fence admiring her. A large number of female soldiers married male soldiers which sometimes caused difficulties for the military as, after all the training and money expended on the women, they often did not complete their first three-year enlistments. Due to the rules of the day, the women were forced to apply to be discharged when they wished to get married. A number of the women who married Vietnam veterans have realised the healing potential in sharing their stories, hearing them heard and knowing that others have also experienced some or all of the difficulties that a number of them have faced.

Weapons training for women was not introduced until the late 1970s for some Citizen Military Forces (CMF) units and in 1985 the WRAAC was disbanded. From that day forward, the women trained with the men, wore the same uniforms and were housed in the same barracks as male soldiers.

I interviewed sixteen women for the *Lady Soldiers* film but this book was a much more time-consuming project than making the film. I would have liked more officers and nurses to have told their stories. I contacted a number of former Army nurses but they did not wish to participate. Several of those who had served in Vietnam said they found it too distressing to relive those times. WRAAC officers also seemed reluctant to share their stories although I am grateful to the ones who have participated, but the majority of women I interviewed were eager and prepared to talk of their lives and did so very candidly.

Recently an Army training establishment contacted me to request the use of the *Lady Soldiers* film as part of their training program. I was very happy to allow them to do this.

Major Cheryl Goulding, the Officer Commanding 10 Transport Squadron in Perth, attended the premiere of the *Lady Soldiers* film in 2012 and emailed me the following comments;

> What an inspirational documentary about the Lady Soldiers of the WRAAC!!! Thank you for … getting these ladies' voices finally heard. I have NEVER seen or heard 'warries' from the female's perspective. It was truly engaging, heart-warming and eye-opening. Thank you again for sharing these precious memories … especially from the women who made a difference for us current serving women. Your efforts have not been in vain, ladies. We are now recognised for our abilities and fortitude, which equates to 'equal pay for equal work'. Praise God I don't have to give up my career to get married and have children. A remarkable documentary portraying women serving our country. PAR ONERI (Equal to the Task).

I believe that these comments show how far women in the military have come in those sixty-plus years since the formation of the WRAAC, and that the women whose stories you will read were pioneers who served our country in peace and war and paved the way for the women who followed them.

I hope that the reader will be inspired by the joie de vivre, camaraderie, resourcefulness and courage of this group of women, many of whom have not had easy lives because of their husbands' service to this country during wartime. I am very grateful to them all for trusting me with their stories and I hope that I have done them justice. I salute them all.

Lyn Kane Dale

the WRAAC

WRAAC SELECTION

To be eligible to join the Women's Royal Australian Army Corps, you must meet the following requirements:–

Age: 17 and under 30 (with previous military service, up to 38 years). If under 21, you must produce written consent of parent or guardian to your enlistment.

Education: A good general education is desirable and of course you must be of good character. If you have served previously, you must have a satisfactory service record.

Nationality: Australian citizen or a naturalised British subject permanently resident in Australia.

Marital Status: Single (or a widow or divorcee without dependent children).

Medical: Medically fit in all respects to Army enlistment standards.

Before being accepted, a WRAAC Officer will discuss with you the training and employment available in the Corps based upon your own preferences, experience and test results. Following selection, you are told when you can commence your basic training course which will be so arranged that you have time to complete your personal arrangements.

PERIOD OF ENLISTMENT

On first joining the WRAAC, you have a choice of enlisting for 3 years or 6 years. Thereafter, you may re-engage for further 3 or 6 year periods.

Chapter 1
Enlisting

> *has a lot to offer.....*
>
> Have you ever stopped to think what kind of life a girl leads in the Women's Royal Australian Army Corps (WRAAC)?—why she joined the Service?—how you would fit into the life? Have you ever thought that, in most ways, you are like her—keen to get on, to do an interesting job, to make the most of your youth and energy?
>
> Whatever you do in the WRAAC is something to stand you in good stead all your life. You'll become one of a team where individuals play a vital part. You'll share the companionship of girls—and men—of your own age, both on and off duty.

Women's stories have often been overlooked, especially those of 'ordinary' women. It is widely known that women served in the Australian Defence Force in World War II but many people are unaware that more than 10,500 women enlisted in the Women's Royal Australian Army Corps (WRAAC) from the time the corps was formed in 1951 to the 1980s when women were absorbed into the mainstream Army. A much smaller group of women during those years enlisted in the Royal Australian Army Nursing Corps (RAANC).

These young women had diverse reasons for enlisting in the Army. Many came from families with military backgrounds and spoke with pride of the service of their fathers and grandfathers in the Australian Army in the First and Second World Wars and Korea. Many later married soldiers and eventually some of their children carried on the family military tradition. Often the attraction was the diversity of women's roles in the military of the time and being able to take on more interesting and less traditional jobs than those available to women in civilian life. There was also the chance of a better future and being able to earn more money than in 'civvy street'. For the majority there were opportunities to travel within Australia and for a small minority the chance to serve overseas. Overall, especially for the women who enlisted in the 1950s and 1960s, the common thread running through their stories is their decision to choose the military as a means to obtain freedom and independence.

In the 1950s and 1960s most young women lived at home until they married. Their lives were in many cases quite restricted, with parents overseeing their daughters' social lives, vetting their friends and often choosing the kind of work their daughters were to engage in. The Army may have seemed restrictive with its rules and regulations but to a great many women who joined up it represented freedom. As long as they followed Army rules they could be in charge of their own lives.

In those days female soldiers did not have the same freedom as male soldiers, being allowed out of barracks only until 10 pm on weeknights and until midnight on weekends unless they had requested a late pass. However, they were free to choose their own friends, mix with members of the opposite sex and drink at a young age if they wished. In some Australian states the legal drinking age was twenty-one but once they were in the Army they could drink at seventeen regardless of where they were posted. Some felt that Army life was like 'being part of a family' where they had the companionship of young women their own age and often made lifelong friendships based on the experiences they had in common.

In late 1963, having just turned eighteen, I applied to join the WRAAC after seeing a WRAAC corporal at a booth in an arcade in Perth promoting the Army. I read the recruiting brochures and the thought of flying off to Sydney, which in those days was as exotic a destination as Paris or New York is today, was so exciting. In due course I was called in for interviews, a medical examination and a psych test. I was informed by letter that I had been accepted, would be sworn in on 27 February 1964 at Swan Barracks in Perth and would be leaving that night to fly to Sydney for basic training.

Very nervous about the step I was about to take, I reported to the barracks and found five other girls who were to enlist and travel with me. Enlistment was for three or six years, though most signed on initially for three. We swore an oath of allegiance to the Queen and while filling in the necessary forms, I wrote my nationality as Australian. The big male, ex-British-Army sergeant in the office looked over my shoulder and said 'You can cross that out'. As I had not yet learned that if a sergeant told you to do something it would be a good idea not to question it, I said 'Why? I am an Australian. I was born here'. He replied, 'Listen girlie, you are a British citizen and you cross out Australian and write British'. When I later told my parents about this my father became angry and said 'I went to fight a war for this country and my daughter can't call herself an Australian! Your grandfather would turn over in his grave'. My Irish Catholic grandfather had left Ireland in the 1880s to escape sectarian discrimination and had no time for British authoritarianism in his new country. After all the paperwork was completed we were given our Army numbers, something which a service person never forgets. We were each issued with a large brown suitcase, a small brown leather attaché case and our airline tickets and told to meet that night at the airport. I then went home to complete my packing. My parents, brother, three sisters, boyfriend and friends gathered at the airport to bid me farewell as I embarked on the big adventure. My life would never be the same again.

Our plane departed at 10 pm and arrived in Melbourne early the next morning. We had breakfast at the airport and then flew on to Sydney, arriving there about 11 am. Having travelled for more than twelve hours with little sleep we were all very tired. Army personnel were at the airport to meet us and ushered us to a green Army bus. We then went to Central Railway Station to pick up several other groups of girls who had arrived by train from interstate. Sydney looked huge to me after sleepy little Perth. The first sight of the Harbour Bridge and the hilly bayside suburbs was beautiful and I looked forward to doing more sightseeing when I could. Up until then it had all seemed like a dream. After we arrived at the barracks which was to be our home for the next six weeks a female Warrant Officer entered the bus and, in a loud voice, told us to pick up our belongings and get off the bus. As we sleepily struggled to collect our things she bawled at us to hurry up. That woke us up to reality!

The following stories, told in their own words, highlight women's unique contribution to Australia's Defence Force and some of the specific issues and challenges that they faced in choosing what was an unusual career path for women in this period.
Lyn Kane Dale

Background

I've got a long association with the Army. I had two grandfathers in the Army in World War I. One of them was in Egypt and also in France. My other grandfather was a bit old for joining the Army. Why he went I don't know. He was thirty-eight and got killed on the Menin Road in Belgium. My father served in World War II. I had an older brother in the Survey Corps and another brother was in the Citizen Military Forces (CMF) so Army was in the blood.
Desiree Harms (enlisted 1965)

My two grandfathers served in World War I. My paternal grandfather, William Harold Webb, 14th Field Artillery Brigade, Australian Imperial Forces (AIF), was a survivor of the Battle of Fromelles [France] in July 1916 and returned safe and sound at the end of the war. My maternal grandfather, Alan Raymond Griffin, also served on the frontline in France in the AIF. He was wounded, hospitalised in England for nine months and then discharged as medically unfit with a disability that he suffered from for the rest of his life. In World War II my mother served in the Women's Auxiliary Australian Air Force (WAAAF) for three years as a Clerk at Defence Headquarters, Melbourne, and my father served with the AIF in Darwin, Morotai (Netherlands Indies) and Borneo. I joined the WRAAC in February 1968 and worked in Signals. My son from my first marriage enlisted in the Army and is now a policeman in the Northern Territory.
Trish Webb (enlisted 1968)

I come from a long line of military people. Apparently we had a general in the Crimean War and my mother told me I had two uncles in the Russian Cavalry. My grandfather on my father's side was in the Royal

Observer Corps in the First World War. He was also a prisoner of war (POW) of the Germans. The Second World War saw my father and his two brothers sign up. Dad was in the Army and John and Dick were in the Royal Air Force (RAF). Dick, flying one of the new Hurricanes, was shot down in Belgium and died a few days later as a result of his burns. My father, a sergeant in the British Cavalry, was in Germany with the occupation forces after the war and that was how he met my mother as the regiment needed single women for the Regimental Ball. She was a Polish refugee who had survived the bombing of Dresden and was an interpreter although she couldn't speak a word of English. She did however, speak five other languages. She later married my father in Germany. I have two older brothers, one who became a squadron leader in the Royal Australian Air Force (RAAF) and the second who was in the Army in bomb disposal until he retired as a major. In 1982 it was my turn. At the same time two cousins of mine were also in the Army. The next generation is my niece and nephew and my eldest son who are currently all in the Army.
Catherine Sanders (enlisted 1982)

My grandfather served in the British Army during World War l and then immigrated to Australia in 1920. He joined the Army (in Australia) for World War II as a Warrant Officer Class 1/Regimental Sergeant Major (WOl/RSM) and his job was training and overseeing the restoration of Fortuna, the building that was to house the Survey Regiment in Bendigo, Victoria. My father was too young for World War II but he joined the CMF and later the (Regular) Army and was part of the training cadre for those going to Korea. He was present at planning meetings held at Fortuna, Bendigo, Victoria, and was able to note the improvements in place since his days spent there during his school holidays with his father. My turn at the connection with Fortuna came through Survey and to top it off my son was Officer Commanding (OC) at the Bendigo CMF depot and was able to inspect Fortuna before it was handed to the Historical Society. Four generations of one service family with some connection to one lovely old mansion.
Hilary Jones (enlisted 1970)

My great grandfather, Major George Priestly, fought in the Crimean War (1853-1856). He had thirteen children and named most of them after soldiers in his regiment, hence all the thirteen had three or four

names. My great uncle Private 335 Henry James Vivian Priestly was killed on ANZAC Day, 25 April 1915. He managed to get off the beaches and up a small hill known as Baby 700 before he was gunned down. His brother, Gunner 31502 Victor Ridgeway Mathew Priestly, was part of the Light Horse Regiment and although he returned in one piece physically he always carried with him the sights and sounds of battle until his death in 1966. My great aunt, Sister Adeline Beatrice Gertrude Priestly, served with the Australian Army Nursing Service on the frontline in mobile hospitals. Beatrice joined the AIF in Perth (WA) aged forty-three years which was older than the thirty-nine years age cut-off, but she had lots of experience as a matron in various hospitals. I have been guest speaker at several local schools sharing what ANZAC Day means to me.
Libby Flemming (enlisted 1969)

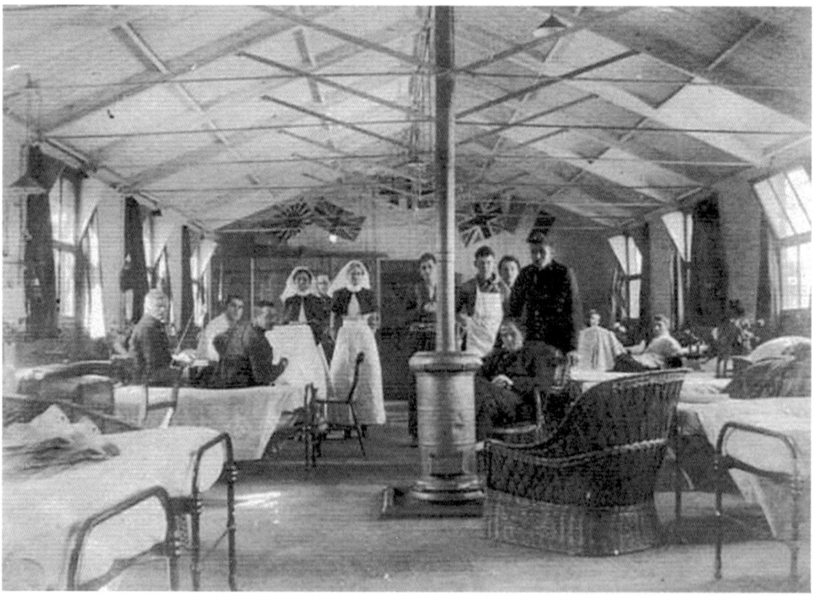

Figure 3 Gertrude Priestley, great aunt of Libby Flemming, with nursing staff and patients in a military hospital in England c.1916

My grandfather was going away to World War I and was leaving from Sydney. My grandmother wanted to see him off so somehow she went from our town, about two to three hours out of Brisbane, all the way to Sydney. She had my uncle as a toddler and was very heavily pregnant with my Dad. While in Sydney Dad decided to be born so his first name became Sydney! Dad was in the Second World War although he did not go overseas. He was a Military Policeman (MP) in Townsville and was also a driving instructor up there. He rather liked this as it was the one time he could tell high-ranking officers what to do. If they did not do what was expected he could fail them.
Daphne Gardner (enlisted 1964)

My father was a very early volunteer at the beginning of World War II as his number was WX347. He was a piper and drummer in the 2/11th City of Perth Regiment and he went to the Middle East with the 6th Division and his battalion went off to Greece and later to Crete. The 2/11 Battalion couldn't be evacuated and he was taken prisoner on Crete and spent four years as a prisoner of war. He was in about three or four camps before the Germans opened up a particular camp, Stalag 383, at a place called Hohenfels near Munich. There were some 5,000 soldiers or so there. This was a camp for Australian Non-Commissioned Officers (NCOs) who refused to work for Hitler, very recalcitrant men who refused to obey German commands or who were constant escapees. The Germans thought this camp was escape-proof but it wasn't in the end
Iris Petrass (enlisted 1961)

My father was a member of the 2/16 Battalion, a West Australian unit, and served for five years in World War II in the Middle East, North Africa and on the Kokoda Trail in New Guinea. He was one of a group of men, most of them wounded, who were cut-off behind enemy lines for ten days as they tried to evade the Japanese and get back to their own people, which they succeeded in doing. Dad had shrapnel wounds and was evacuated, put on a hospital ship and sent to Queensland where he spent three months in hospital recovering. He was later invalided out of the Army. After reading accounts of the fierce fighting in New Guinea I believe he was very lucky to survive and I am very lucky to have been born. Initially Dad didn't want me to enlist as he didn't think I would like the Army, but once he got used to the idea he was very proud of me being in the military. He wrote to me several days after my departure from

Perth for basic training in Sydney, to say that he never seemed to find an opportunity to have a private chat with me before I left but he wanted me to know:

> I felt so proud of you as you walked out to the plane, with your head up, looking so brave and all the time I knew how hard it was to do. For you see, Lynette, I have had to do it so many times myself and it never gets any easier. But now you have settled into your new surroundings I am sure you will do well. Just obey orders and do the best you can and those over you will soon realise your worth. Above all, whilst you are away you should take every opportunity to see everything there is to see in and around your depot. It is an education just to travel around with your eyes open. One can learn so much by seeing it at first hand… Go out of the camp when you get the chance. It helps you to relax and you feel more like facing another day of army routine. I had five years of it and I know. Remember, wherever you are, your mother and I are thinking of you and praying for you too.

Lyn Kane (enlisted 1964)

At the age of nineteen, along with my sister, Marilyn, I enlisted in Hobart, the year of the 1967 Hobart bushfires. We decided to join the Army because the idea of travel and the wide range of opportunities to do something different appealed to us. I guess we chose the Army as our Dad had been in the Second World War and was always telling us stories of Army life. My employment prior to this had been as a clerk in the advertising department of a local newspaper.

Andrea Woods (enlisted 1967)

During World War II Margaret Elder sent away for a recruiting booklet for the Air Force. They kept writing asking her to enlist and her mother had to get in touch with them to say her daughter was only ten years old.

I thought I wanted to join the Air Force because I used to see these beautiful girls in uniform in magazines. I thought it would be wonderful to grow up and look like one of these girls. The war was on at this time and Sydney was full of Service people. The men looked so handsome and all the girls were so beautiful. I'm not sure what they did with the girls that weren't very pretty but all the ones that I saw seemed to be really lovely. Mothers had a lot of influence on girls in those days. I was living in Sydney with my aunty and my mother was in the country. I wrote to my mother and said that I was thinking of joining the Army and she wrote back and told me that I wasn't allowed to in no uncertain terms.

Then in the early 1950s they started to form the CMF for girls and when I was twenty-one I joined the CMF.
Margaret Elder (enlisted 1953)

I joined the Army, the CMF initially, in Sydney in 1961 aged twenty-six. It was during the Cold War period, before Vietnam of course. I felt one needed to be ready to defend one's country in case something might happen. I think I was a bit pessimistic in that regard. I was at work one day and saw an advertisement in the *Sydney Morning Herald* about joining the WRAAC CMF and I applied. A week or so later I arrived at the barracks for an interview, then the psych test and medical, followed by signing on the dotted line. In those days the CMF joined for only one year at a time. I had been in the CMF for six years before I transferred to the Regular Army. As a child I went to boarding school so I was rather institutionalised anyway and I had no worries about living with other people or obeying commands. Life was difficult at times but you put up with those things. The camaraderie overall was the big winner.
Iris Petrass (enlisted 1961)

The idea of joining the Army crossed my mind as a 12-year-old, as two of my father's sisters had been servicewomen in World War II and my father was a returned soldier. Employment opportunities in my home town of Robinvale were quite limited. My parents were wanting a career that they thought would benefit me, and the Army sounded to them like a very good option, plus on reflection, I have no doubt that they thought a bit of discipline would do me the world of good - and it did. Thanks Mum and Dad. Although it was not a burning desire for me, they insisted I send away for the application forms which I duly did on my seventeenth birthday. Lo and behold two months later, in January 1966, I was in Sydney and learning to march, spit polish my shoes and hone my ironing skills plus the many other requirements of a recruit.
Lyn Morris (enlisted 1966)

I had always wanted to join the Forces from an early age. Maybe it came from my father's involvement in World War II. Before enlistment, I worked at the Brisbane Telegraph as a Teleprinter and Telex operator in the communication rooms. I sent news items to other states, and received items in return plus overseas news through Australian Associated Press (AAP). I joined at the tender age of eighteen. I remember Dad at the

Brisbane Airport, waving with tears running down his face. He was fiercely proud of my achievements.
Denise Darbey (enlisted 1973)

Figure 4 Recruits in 1964. Lyn Kane (center back row) and Chris Cain (kneeling front right)

A couple of months after my eighteenth birthday, my mother suggested I enlist – I was getting pretty restless at home – and I became the fourth generation in my mother's family to join the Army, including my mother who was in the Australian Women's Army Service (AWAS) during World War II. My father had also served in Darwin during the bombing so it wasn't too big a step for them. I was quite overwhelmed with the whole procedure but excited to arrive at WRAAC School for my recruit course.
Kay Owen (enlisted 1964)

I was born in 1955, in Adelaide, South Australia, the youngest and third child to migrants who came to Australia under the Displaced Persons Resettlement Scheme, arriving in 1950 from Germany. My father was from a tiny village in Poland and my mother from a tiny village in Germany. During World War II my father, a soldier, was captured by the Germans and held as a prisoner of war. At the end of the war, upon release, he could not safely return to his home, as soldiers who fought against the Germans risked being shot on re-entering Poland or forced into a gulag. My father stayed in Germany, doing unskilled work, and there he met my mother. After arriving in Australia my parents were sent to Woodside Migrant Hostel. My father worked as a semi-skilled operator at British Tube Mills in Adelaide and my mother worked as an unskilled labourer at Simpson Pope (white goods). Both worked to pay off a house in Prospect, an inner city suburb in Adelaide. It was not a happy marriage. I remember a lot of arguments and things being thrown around the house, with holes in walls and doors. I was ten years of age when my father died. My memories of him are scattered (because) as a latchkey child to hard-working immigrants who worked shiftwork, I did not spend a lot of time with him. After he died, the fighting continued, but between my mother and her children. The trauma of war had obviously seriously affected my father and mother and as a consequence, us, the children. My mother took on numerous part-time jobs, which meant that I spent most of my time after school cooking and cleaning up after my two older brothers and myself. Participating in after-school sport activities was not considered important. My mother had no family support in Australia. I usually only saw my mother for a few hours on weekends, as she was always working, trying to provide for us. It was a very lonely childhood. My oldest brother was a bit of a wild one, so he was hardly around and my middle brother, the extreme opposite, very introverted and a loner, spent most of his time behind closed doors. The only 'joy' I remember was going to my local Catholic Church, where I often attended Mass numerous times during the week, mostly alone, except some Sundays when I attended with the rest of the family. My mother suffered from what we now know as Obsessive Compulsive Disorder, demanding extreme neatness and cleanliness, to a point of punishment if her standards were not met. Our family was dysfunctional. There was regular verbal and physical abuse. During high school I had to complete a series of tests - I still do not know why – however, the results necessitated the school to call my mother in for a meeting. I thought I

was in trouble but the teacher wanted to encourage my mother to change the class she had me enrolled in, from Commercial to General Track, which would ensure that I could enter university at the end of Leaving Honours. My mother did not change her mind, stating, 'My daughter will get married and have children and I need her to get a job to help me', plus 'Nice girls work in offices'. I completed the Commercial Track, securing a job as a stenographer with a city-based insurance company. I enlisted in 1973 in Adelaide, three days short of my eighteenth birthday. As I was under-age, I needed a parent to sign to give permission. My mother agreed to sign, but only after I agreed to allot her $50 per month from my Army pay. This was the amount I was paying for board when living at home. I did not begrudge this demand, as I knew she had worked very hard, and had had a tough time raising the three of us, plus my wages during recruit training were almost double what I was earning in my civilian job. I had this amount of $50 allocated from my salary every month until I married in 1975. I was so controlled by my mother, always wanting to please her, listening to her disapproving voice and ongoing put-downs that I needed to break away, find out who I was and make some happy memories.
Maria Adamczak (enlisted 1973)

After finishing school I found a casual job at a factory. I'd also done a receptionist course because I hadn't passed my High School Certificate (HSC). I'd gone for an interview to work as a bank teller and they said, 'Oh no, sorry, you're too old. You should have come and seen us when you were in Year 10', so I was in limbo with regard to what to do with my life. I read a newspaper article in the *Herald* newspaper saying that the Army was increasing the numbers that they were taking in. I'd entertained the thought of applying for the Navy as a radar plotter until I saw the article so I applied for the Army. I went to WRAAC School in September 1976, a week shy of my nineteenth birthday. I was glad to leave home, to be quite honest. I was from an ethnic Italian background and as much as Mum wanted me to finish school, the attitude was 'You don't have time for friends. You don't have time to read or go out. Learn to knit and learn how to use a broom because you're eventually going to get married and you won't need all of that stuff'. Oh it was horrid and plus I was the eldest of four girls. I was bringing up my younger sisters. My youngest sister was eighteen months old when I joined up so I was more or less like their mum. I was glad to get out of it. My second sister, who's only fifteen

months younger than me, resented my having got away. She was still at home until she was twenty-four years of age and by then I'd got married. Even though mum wasn't happy about my leaving she didn't try and stop me because I said to her, 'Look, I'm eighteen. I don't need your signature. I'm just letting you know and I'm doing this as a courtesy'. My father said, 'It's her decision. She's got a three-year commitment. If she doesn't like it she's got to stick it out for that time'. Because Dad knew of the Italian Army being as strict as they were he thought that the Australian Army was along the same lines. That you didn't get much liberty. Well nothing was further from the truth. So with Dad having spoken that way Mum didn't push the point. But, as the years wore on, Mum took a lot of pride. To this day she still boasts about my having been in the Army and the rank that I've attained and all that I've achieved. Being foreign, Mum didn't want to see me go, but, whenever I came home, she always used to say to me, 'Now make sure you come home in your uniform'. She wanted to show me off. She became very, very proud, not that she understood fully what it entailed, but nonetheless she was happy that I chose that career path and that it worked out for the best.
Anna Celi (enlisted 1976)

I was working on a telephone exchange in the country town of Alexandra in Victoria, a pretty quiet place. I was looking through the *Women's Weekly* and there was an advertisement for the WRAAC. The uniform looked so good. The list of things that we could do included cook, driver, draughtswoman, storewoman and clerical. I immediately thought I wanted to be a driver, as all I ever wanted to do was drive a truck.
Betty Hall (enlisted 1963)

I grew up in Texas, Queensland, where my father was bookkeeper for Texas Station, a 65,000 acre property situated on the border of Queensland and New South Wales on the Dumaresq River. I had an idyllic childhood riding horses, milking cows, tending chooks and helping my dad with the sheep kill each fortnight for our meat supply. I was the youngest of four girls and my dad's right arm man. As he always said, he never got a burr cutter or a boundary rider amongst us. I attended the convent school until Grade

Six aged twelve with the Sisters of St Joseph teaching, or rather caning, us into learning mainly by rote. At the completion of Junior Certificate, which is Grade 8, the only option was boarding school in Toowoomba or Brisbane. My two elder sisters had been at boarding school for several years and hated every minute of it. My parents decided to move to Tasmania in December 1963. My sisters all became teachers as my mother decided that all her girls would be teachers or nurses. I didn't want either so I decided to find out about the Army. I had an interview and psych test in Hobart, 350 kilometres from home, then waited to hear if I'd been successful. I had returned to Devonport High School to start my Year 12 but gladly packed it in when I was accepted into the WRAAC. At school they could never work out whether you were an adult or a child at seventeen, but the minute we set foot on WRAAC School soil we were expected to be adults and I liked that. I received notification that I was to report to Anglesea Barracks in Hobart to be sworn in. Private Flemming was born. The adventure began.
Libby Flemming (enlisted 1969)

I had to travel from Launceston to Hobart by train to enlist at Anglesea Barracks, staying at a hotel overnight before flying out to Sydney next morning. That was in 1962 and I was eighteen years of age. Leaving was very hard. I remember phoning home and sobbing for hours as I sat in that hotel room scared (of what?) and all alone thinking 'WHAT AM I DOING?'.
Julie Ross (enlisted 1962)

I decided to join the CMF as most of my girlfriends were going steady or were engaged and I thought 'There must be more to life than that'.
Liz Stone (enlisted 1967)

Advice from my eldest brother who was a serving Army member at the time to his 17-year-old sister about to embark on her life-changing adventure:

Brother: 'If you join the Army I won't speak to you again.'

Cherryl: 'Why not?'

Brother: 'Cause all Army girls are Officers' mattresses.'

Cherryl: Well!! I did join the Army and he did talk to me and no, I didn't get to sleep with any officers.
Cherryl Burrows (enlisted 1964)

My mother's expectations of me were nothing short of controlling and strict. There was a German saying my mother had that translated in English to: 'Army girls are Officers' mattresses'. My mother made it very clear that I was not to become one of those. There was never any doubt in my mind that I would not be a virgin for my husband.
Maria Adamczak (enlisted 1973)

When I was eighteen I joined a bank – the National Australasia in Brisbane. At age twenty-six I set off to see Europe. It was the kind of thing you did about that time. I travelled alone and had six wonderful months discovering the sights, sounds and smells of foreign countries. My position in the bank had been held for me, yet I was restless and looking for new adventures. Then I saw an advertisement in the paper for the CMF. I had absolutely no idea of what being in the Army would be like.
Bethmay Watson (enlisted 1975)

I decided to join the Army because it was a chance to do something different. When I was a child I noticed that when Dad came in from milking with Mum, he read the paper, which we were not allowed to touch and Mum had to keep on working and cook the tea. He was a good man BUT, I said to myself, 'When I grow up **I** am going to get the paper'. The times they were changing and I did not want to be a wife and mother but to be me and have a life for me. I enlisted in Brisbane in 1967 when I was twenty-two and a double-certificated nurse - general and midwifery. I feel I adjusted well; I was posted to Ingleburn, NSW, 2 Military Hospital, the largest military hospital where all the casualties from Vietnam came to start their recovery. The work was interesting - plenty of parties and close to Sydney where I had been living the previous year.
Ann Hall (enlisted 1967)

In 1964, when I was seventeen years of age, I enlisted in Brisbane. I had been an assistant nurse for two years and wanted something different when my brother brought home brochures for all the Services. The information on the Army intrigued me and, after talking to Mum and Dad, I set the wheels in motion to enlist.
Daphne Gardner (enlisted 1964)

HOW TO ENLIST

No leaflet can possibly explain all aspects of life in the WRAAC nor answer all the questions you may have. If you would like to know more, the WRAAC adviser at your local recruiting centre will be pleased to talk it over with you and answer any queries you or your parents may have.

Call, at your own convenience, at one of the addresses shown below and learn what life in the WRAAC can offer you. Discuss with the staff, your preference for training, employment and posting, this could be your opportunity to lead a life with a difference.

QUEENSLAND

Army Careers Officer,
130 Mary Street,
BRISBANE.
QLD., 4000.
Telephone 31 1031

Army Careers Officer,
52-60 Sturt Street,
TOWNSVILLE.
QLD., 4810.
Telephone 71 6784

NEW SOUTH WALES

Army Careers Officer,
18-20 York Street,
SYDNEY.
N.S.W., 2000.
Telephone 29 2122

Army Careers Officer,
809 Hunter Street West,
NEWCASTLE, 2300.
Telephone 61 1063

VICTORIA

Army Careers Officer,
301 Flinders Lane,
MELBOURNE.
VIC., 3000.
Telephone 61 3731

Army Careers Officer,
7 Exchange Arcade,
GEELONG, 3220.
Telephone 96 780

SOUTH AUSTRALIA

Army Careers Officer,
125-127 Pirie Street,
ADELAIDE.
S.A., 5000.
Telephone 23 2891

WESTERN AUSTRALIA

Army Careers Officer,
Commonwealth Bank Building,
55 William Street,
PERTH, 6000.
Telephone 23 1356

TASMANIA

Army Careers Officer,
Anglesea Barracks,
Davey Street,
HOBART.
TAS., 7002.
Telephone 20 0141

Army Careers Officer,
Paterson Barracks,
St. Johns Street,
LAUNCESTON, 7250.
Telephone 31 1005

I think that I was a foolish young girl who didn't have a clear idea of what I might like to do and somehow got the impressions that it would be 'fun' to join the Army. Possibly hearing about the male/female ratio might have assisted the decision. I knew quite a number of young men who'd joined up and my family has always been involved with the Royal Australian Navy (RAN) or the Australian Army. Before enlisting in Sydney in 1966 at the age of eighteen years I did a stint as a nanny. That was a lost cause, although I did learn to ski which was a positive. I then attempted to do a hairdressing apprenticeship but fairly quickly discovered that a) I had no skill and b) I had no interest in standing around washing hair and inhaling hairspray for a living.
Kim Lake (enlisted 1966)

I was exempted from school at age thirteen in order to look after my younger siblings because we had lost our mother. Later on I worked in the Calvary Hospital in North Adelaide as a kitchen hand. I wanted to be a nurse so I studied and passed my nurses' entrance exam, but I enlisted in the Royal Australian Army Nursing Corps. I sat for the Army test sometime in July 1960 and was surprised to learn that I was the only one to get through to the medical and passed that as well.
Gloria Coombe (enlisted 1960)

I grew up in Aberdeen, NSW, a small country town, and always had a dream of moving away and doing something different but really didn't know what. At that time I had two cousins in Vietnam and we were very isolated from the negative press about (the war). An ex-WRAAC, Jan Devine, came to work at our office and she spoke very fondly about her time in the Army. I did some research about the Army life for a young girl and felt it was for me. I am the eldest of seven and had a very strict Catholic upbringing so I knew I could cope with the discipline. I enlisted in Sydney, in May 1972, and had my nineteenth birthday at WRAAC School.
Fran Sinclair (enlisted 1972)

It was 1976 and I was in Sydney reading an article in *Cleo* magazine about women doing different jobs. One was an instructor at the Women's Royal Australian Army Corps School in George's Heights (Mosman, Sydney). She seemed to have a very interesting job teaching officer cadets how to conduct themselves in the Army and it appealed to me because I

was a bit of an outdoorsy sort of girl and I'd been looking for a new challenge. I thought, 'Maybe the Army is for me'. On day one at WRAAC School the lady in charge was Captain Cheryl Campbell and she said to me, 'Tell me why you joined the Army'. I said, 'It was you in the article in *Cleo* magazine'.
Denise Dunn (enlisted 1976)

I joined the Army at seventeen - the only career or occupation that I considered at any time. My mother wasn't very happy with the idea so I forged her signature on the enlistment papers and then presented her with the fait accompli.
Sheila Cusack (enlisted 1964)

I had a reasonably good job as a telephonist in Kojonup (WA) in the South West. Everyone was into marrying and there was no future for anybody there. I enlisted in 1960 in New South Wales.
Julie Elverd (enlisted 1960)

My enlistment was from July 1955 to December 1958. I enlisted in NSW at eighteen years of age and left for WRAAC School, Queenscliffe (Victoria) two days later. It was my first experience of interstate and the rickety old NSW overnight train to Melbourne.
Wilma Stralow (enlisted 1955)

I was born in country Queensland, the sixth girl of a family of six girls, so I was the baby. In 1959 when I was thirteen my mother died. I went to work in Monto doing various jobs. I was a waitress, a barmaid for some time, and then I got a job at the local hospital as a domestic. My father started writing to a lady in Germany and he sold the family home and went to Germany to meet his future wife. I moved in with my sister and her family. One of my school friends who had joined the Army was home on leave and mentioned that I should join because she loved it. I sent off and got the brochures from Brisbane and they looked wonderful - all the glossy pictures, etc. - and I decided to join the Army nursing corps because I guess I was familiar with the hospitals.
Ruth Grace (enlisted 1964)

I joined the Army in June 1965 in Adelaide, South Australia. I had tried to sign up just after turning seventeen years of age, but because I

wanted to join the RAANC I had to wait until I was eighteen. I'm not really sure of their reasoning behind that but girls had to be eighteen to join the Nursing Corps. A lot of my friends had day-to-day mundane type jobs and were in relationships that weren't very satisfactory and I thought, 'No, I'm going off to do something totally different'.
Marion Quick (enlisted 1965)

My Dad was in the Army. I was bored as a legal secretary and wanted a bit more adventure. I used to lodge legal documents right next door to the Army recruiting office and one day in May 1980 I went inside and decided to apply to join. Initially I was told I wasn't accepted - not sure why. The sergeant at the recruiting office came knocking on my door and said he had been fighting for two days to have me accepted. I was a bit scared at that, thinking I would be up against much smarter girls, people with a lot more skills or education. When I got there, I was a little insulted because I wondered why they had rejected me in the first place.
Sandy Jarrett (enlisted 1980)

At the age of nine or ten my parents broke up and my sister and I went to boarding school for about twelve months. I then went to live with my mother in Parkes, NSW, where I finished high school, then to Sydney to live with my aunt whilst I did a secretarial course which I hated.

When Jenny was sixteen her mother died.

When we went to my aunt's place there were always conversations about who was going to pay for our schooling, our clothing and so on. I'm sure the family didn't realise that little kids have big ears and so we heard all these things. As a consequence of that I wanted to make myself self-sufficient as soon as I possibly could, so when I saw the ad to join the Army I thought, 'This sounds like me'. I rocked on up and was accepted. Consequently joined up in August 1962.
Jenny McLay (enlisted 1962)

I worked in a shoe shop for a short while but left after just one too many not so subtle 'brushes' past from the boss and too much holding of the ladder whilst I was at the top retrieving yet another pair of shoes. I found a job in a shoe factory where I stayed until I joined the Army. My initial enquiry was for a dental nursing position but when I found you had to be eighteen (to join the nursing corps) I swapped allegiance and

applied to the WRAAC as a driver. I never considered any other job. I travelled to Sydney in September 1968 for my tests and was accommodated overnight at a hotel in Kings Cross! It was all very grown-up and worldly for a country girl who had never done anything without her family. On the day of the psych test I befriended another girl and we were two of only three girls remaining at close of business. Her name was Maree Alewood and she lived in Sydney. We promised to keep in touch and let each other know how we got on. We both enlisted in Sydney on 17 January 1969. I was seventeen and a half.
Deborah Harding (enlisted 1969)

I first thought about joining the Army when my twin sister asked me to go with her for an interview. Until then it had never crossed my mind. It was 1964 and I had just turned eighteen. As I had started my nursing training at the Royal Brisbane Hospital, it was suggested that I should join the Royal Australian Army Nursing Corps, which I did in June 1964. I had no idea what was in store for me. I left Brisbane with five other girls and arrived in Sydney to begin training at WRAAC School. The time there was challenging but also enjoyable. I learnt what it meant to be 'on parade', many friends were made and fun times such as the secret (we thought) pyjama party held in one of the rooms. I think the instructors turned a blind eye!! We were fitted out with uniforms and were very proud the day we marched out to our first postings.
Elizabeth Anderson (enlisted 1964)

When I was a child, my parents and I used to visit my older married sister in Sydney's west, and Dad used to take a shortcut through the Army camp at Moorebank near Ingleburn, NSW. I loved driving through the camp and was fascinated with all aspects of it, including the men in uniform and what I could see of the camp and the equipment. When I was old enough to join the Army I had to go to Sydney to have a physical and psych test, and was put up overnight by the Army in the YWCA, having travelled to Sydney by train from Camden alone - a great adventure for me in those days. I remember the psych chap telling me that I was very young and immature, and suggesting that I join the CMF for six months and then re-apply. I pleaded with him to accept me, as I wasn't able to join the CMF, because I lived too far away to travel to the depot. I obviously convinced him, for I received a letter telling me that I was successful. I had been working in a bank in Camden for two years as

a ledger machinist, my first proper job since completing a secretarial course in Sydney. I was eager to enlist, and did so just a few months after my father passed away. It was a big step for a very immature 17-year-old. I remember being met at Central Station by a WRAAC driver, who drove me to WRAAC School. It was one of the most exciting days of my life up until that point.
Marie Taylor (enlisted 1962)

I was eighteen and had a part-time waitressing job in a small country town in New South Wales. With no career prospects, and wanting something more out of life, I called the local Army recruitment agency and after meeting with the recruiting sergeant, pretty much decided that this was something I could do that would be worthwhile. Of course my parents looked on it as 'a phase' and kept telling me that I would never cope with the discipline. I think they were quite surprised when I thrived in that environment.
Clare Williams (enlisted 1980)

I was the eldest of four children and we were living in Shepparton (Vic.) when my mother decided that it was time I left home. I had completed my Leaving Certificate and was working at a corner store. Her answer to my employment was to take me over to Mooroopna to the training hospital to enrol as a nurse. On our way home from an interview with the matron I saw the Army recruiting van in Shepparton. After obtaining information and forms I went home and my father (ex-military) signed them for me. The outcome – I was one of three selected out of forty-eight on the day we sat the test, and I was called up ten days before having to front to the hospital to start training! I consider myself very lucky. I was nineteen and my enlistment was on 12 May 1970. Five years and one hundred days later I resigned.
Hilary Jones (enlisted 1970)

After leaving school I put myself through one term at Business College so I could get a better job. I had been working in an office of a small building company for a year and found it quite dull so I applied for three jobs and they all came up at the same time. Two were with local councils and the other with the Army. I chose the Army as it offered a broader scope of opportunities as well as a flight to Sydney. I had always wondered what it would be like to be in a plane. I joined in Adelaide in

1975 at the age of seventeen and a half. In the first year my mother would ring the Army and complain about my lack of contact. I would then be fronted to my boss and told to write to my family. It became easier to keep in contact when my parents got a telephone.
Christina Morgan (enlisted 1975)

I enlisted in Melbourne in 1967 at eighteen years old. I had been a trainee nurse at Royal Melbourne Hospital (RMH), not lonely but restless, constantly looking for challenges and tired of living on twenty-eight dollars a fortnight. For some reason I thought I could train as a nurse in the Army and go to Vietnam. No straight answer from Recruiting. It was only later that I realised the Registered Nurses (RNs) with trauma and accident & emergency (A&E) experience were the only ones who went to Vietnam. They did send a few Enrolled Nurses (ENs) to do the transfers home to Australia on the Hercules (aircraft). I served from 1967 to 1974 and adjusted well to Army life. I had been to boarding school so was ready for it. Everything was always interesting and exciting – never a dull minute – even the girls' nights in.
Joie Norton-Willson Scheinpflug (enlisted 1967)

I joined up in WA in October of 1963 and I suppose it was the excitement, the travel and the adventure.
Christine Negus (enlisted 1963)

I was a country kid, who had dropped out of school and was working in a country clothing shop to save some money to travel overseas. I was very conscious that most of my friends had gone on to training courses. Army, Navy, Air Force - six years, four years, three years. Which one was three years? Army. Well that will do. Such in-depth decision-making. I enlisted in Melbourne a month after I turned eighteen and served for eight years. I was a very conforming young recruit and rapidly adjusted to the WRAAC institutionalised environment which was interesting to say the least. I loved the camaraderie of barrack and mess living. Of interest, we were 1,200 girls and 28,000+ men in the Army. Not something I had considered when I applied to join the WRAAC. I really had no idea what to expect. I hadn't considered conscription and the pros or cons about Vietnam.
Jayne Rickard (enlisted 1969)

On the day of my fifteenth birthday I left school and secured a job in a newsagency/post office. Moving around from school to school, I had to repeat and lost interest in my schooling. There were limited opportunities in a small country town. When I read the glossy leaflets about the Army I was immediately impressed with what it had to offer - security, meals and board, and above all training in an occupation. It was not an easy task to persuade my dad to sign the papers. He was convinced I would not pass the test. Well, there were only two girls to get accepted out of forty, and yes I was one of them.
Ann Dullaart (enlisted 1966)

I was a small town Perth girl who was only earning six pounds a week - the equivalent of twelve dollars now. Out of that I had to pay four pounds board to my parents. I had left school at the age of fourteen because the family needed my contribution. I obtained work in a factory making potato chips. I rode my bike to work each day. In those days I had no hope of obtaining further study and no hope of travelling because of the huge cost of travel out of WA. There weren't ample opportunities for low income families in those years so I really wanted to join the Army. Dad forbade me to join either the navy or air force because, in his opinion, in the advent of war they would be the first to be targeted by an enemy. He still held strong memories of World War II. In 1963 there were limited Army intake numbers allocated to WA and I was determined to be one of them. With many girls vying for those few vacancies and being of limited schooling and only nearly five foot two inches in height I made sure I stood as tall as I could possibly get. I arched my insteps till it hurt. How wonderful it felt to join the Army, travel and get training at Georges Heights, NSW, where from the parade ground we watched flying boats coming in to land on the harbour.
Helen McLeod (enlisted 1963)

I was a country girl and hadn't really been anywhere or done anything, so at the age of nineteen I decided this was the way to do it-by joining up.
Esma Summerton (enlisted 1955)

Even at seventeen years old I knew that I wanted to travel and do something with my life. The Army was the perfect solution, to travel, meet people, and always have a bed to sleep in and a good meal at the end of each day. I also think this made the decision a little easier for my

parents when it came to signing the consent forms. I found out about women being in the Army from a couple of girls I worked with. From there it was persuading my parents to sign the consent forms and then came the medical, psych tests and finally the letter of acceptance. I remember that it was a cold June night when we left. People really dressed up in those days when travelling. I wore a red suit with black trim and black shoes. There were a lot of tears when saying goodbye at Perth airport but also the excitement of actually going to Sydney. A totally new life awaited.
Cherryl Burrows (enlisted 1964)

Figure 5 Letter and postscript to Cherryl Burrows from her parents just after she left for WRAAC School in 1964. Her dad was away working and unable to farewell her in person. *(facing page)*

My Darling Cherryl.

I don't want you to
leave home without a few lines from
your Dad. I am very sorry that I
cannot see you off but I promise
never to neglect our darling by going
away + leaving her all the responsi-
bilities in her capable hands so
when you return on leave all the
family will be waiting with open
arms to welcome our travellers. Save
all you can for doe is always necessary
but not by neglecting yourself; please write
home often + share your worries with us
thereby lightening them. Don't neglect your
health + always seek advice if anything
goes wrong. Always remember to Mum + Dad
our children will always be little so please
let us spoil you + expect your confidences.
Help Brian by being Kind + writing, he
really loves you.

Always lovely daughter
Your ever loving Dad

x x x x x x x x x

Didn't expect me to put a bit more in did you? in case
I didn't tell you before I love you oh so deepey darling
hurry up and come home to me.

INTRODUCTION TO THE WRAAC

Your first introduction to the WRAAC will be a five week initial training course at WRAAC School, George's Heights, Sydney. This will be your settling-in period. You will receive an introduction to service life, gain an understanding of the Army generally and the duties and responsibilities of the WRAAC in particular.

You will be fitted with your uniforms soon after you arrive at the School. The main items of summer and winter uniforms are tailored for you by firms in the city of Sydney. In addition you receive a cash allowance of $42 with which to buy your own choice of lingerie and night attire.

Chapter 2

Training

Where you train

The Regular goes first to the WRAAC School at George's Heights, Sydney, for five weeks settling down and learning about the Army and her new colleagues. Adjustment to new routine is closely studied as a guide to assessing her abilities and choice of work. The Army's most keen WRAAC girls have work they suit and like. Here is a chance to "find" yourself and uncover latent talents.

You feel different when you leave the Training School—more alive, self-reliant, ready to tackle with zest the next stage in your career. You can see the ladder of promotion, from lance-corporal to colonel. Anyone with a sense of service and ability can climb it, and will be encouraged at every rung, if she is a trier.

They say that in the Army the food is very fine
But how the hell do they know, they haven't tasted mine
Oh I don't want no more of Army life
Oh, how I wanna go home
Gee but I wanna go, but they won't let me go
Gee but I wanna go home

Verse from Army song

The home for new recruits for the first few weeks of their Army service was the Women's Royal Australian Army Corps School, better known as WRAAC School, situated on the cliffs at Georges Heights, Mosman, overlooking the Heads at the entrance to Sydney Harbour. There were magnificent views, especially from the parade ground, which was on the highest point. While they were drilling on the parade ground, the recruits could see ships making their way through the Heads and up the harbour towards Circular Quay. In the early 1950s female soldiers did their recruit training in Victoria until 1958 when WRAAC School was re-established at Georges Heights.

WRAAC School was a collection of tin-roofed fibro buildings with four girls sharing each room. The rooms opened onto an open verandah and there were large ablution blocks between the buildings. The women's first stop after depositing luggage in their room was the quartermaster's store, known as the Q Store. They were issued with temporary uniforms – beige linen dresses for WRAAC personnel and grey dresses for RAANC - which buttoned all the way down the front; three pairs of lace-up, fractional fitting shoes; a handbag; beret; two shirts; a cardigan and other items of clothing and kit. In 1964 female personnel were paid an allowance of eighteen pounds ($36) to buy underwear and sleepwear. Everyone was issued with bedding – two sheets and a pillow case - which were changed weekly, and four Army blankets. They had their first meal in the Other Ranks (OR) mess and then went back to the barracks to be shown how to make beds Army style – corners of sheets and blankets folded with a knife edge and then tucked in very precisely. The blankets had to be pulled very tightly across the bed so that the top blanket was completely flat with not even a hint of a dip in the middle. The décor in each room was very spartan with four iron beds, four metal wardrobes, four metal desks and four chairs. The floors were brown linoleum which had to be polished every day to a very high sheen. Rooms had to be in inspection order every weekday until 1700 hours (5 pm) so there was no collapsing on the beds at lunchtime for a rest. Many times after a hot, tiring morning on the parade ground weary recruits would go to their rooms after lunch and lie down on the floor for ten minutes' rest before going back to practise marching on a broad front, about turns and doing it all in strictly military fashion.

Many hours were spent on the parade ground during the training. Blistered feet, and in the summer, sunburnt faces were an occupational hazard. Although recruits weren't on the parade ground in the middle of the day it was often very hot and berets did not protect faces from the sun. A number of girls fainted while standing at attention. The first instinct of the person or persons near that individual would be to catch her before she fell, but recruits were told not to do that. If someone started swaying, those near them just had to let them fall; and on a number of occasions the squads marched off the parade ground, stepping over the limp body on the ground. Staff would then come and take the soldier to the Regimental Aid Post (RAP).

Most new recruits had never lived away from home before. The average age was eighteen years and most of them missed home and family while trying to come to grips with the idiosyncrasies of Army life. Most days they would get out of bed at 6 am, shower, dress, line up for breakfast, wash their dishes, go back to the barracks and get rooms in inspection order. If rostered for other duties like bathrooms, the person on bathroom fatigue would have to wait until the last person had finished in the showers and toilets. She would then clean the area until it was spotless, then rush back to her room and change into a clean, starched uniform for inspection.

The uniforms were so stiffly starched they would stand up by themselves after being taken off the clothesline. It would take an average of an hour to iron each one. The uniforms would be dampened down with water, rolled in a towel for a few minutes to be sure that each centimetre of linen was wet and then be placed inside out on the ironing board. Each seam would be ironed flat and then the uniform would be turned right side out and meticulously ironed without a crease anywhere. Each girl would then carefully carry the dress back to her room, put it on and then climb onto a chair. A room mate would button it up taking care not to crease it in any way. The girl would then jump off the chair onto the ground and carefully walk sideways up the steps to the parade ground for uniform inspection, again taking care not to crease the uniform. This happened every day, and because there were not enough irons for all the recruits they rostered themselves to get up in the night to have a turn to use the irons. Underarms got very chafed from swinging arms while marching in such stiffly starched uniforms and feet got badly blistered from marching and drilling in new shoes. Some of the 'old hands' advised that dabbing methylated spirits on blisters would dry them up and eventually the recruits' feet got harder or more used to the unaccustomed lace-up shoes which were surprisingly comfortable once they were broken in.

The women gradually improved their marching, saluting and following of the shouted commands. After giving them a few days to learn the drill, they were marched off the parade ground every time a mistake was made then marched back on again and told to do it properly. They learnt fairly quickly. They spent hours in the classroom struggling with unfamiliar subjects like Army law and procedures; how to recognise people's rank by the stripes, badges or 'pips' they wore; what was expected of them as soldiers, and so on. Every session on the parade ground was gearing them towards their Passing Out Parade[1] at the end of the course followed by postings to their new units. They were issued with pay books and learned that in order to collect their pay they would have to go on pay parade once a fortnight. They were lined up in order of states beginning with Queensland, whose Army numbers began with '1' and so on, to the Northern Territory whose numbers began with '7'. Females had the letter 'F' in front of their numbers. Each soldier would stand at ease and as everyone moved up the line they would come to attention, march one step forward and stand at ease. This would continue until they arrived in front of the officer who was paying them. Each person would come to attention, salute the officer, sign for her pay, hold her right hand out for the pay book, transfer it to the left hand, salute again, about turn and march out of the building. This continued every payday for the rest of their Army careers.

There was a canteen at WRAAC School which sold a little of everything – soft drinks, potato chips, chocolates, cigarettes, and also cameras, electric irons, transistor radios, stationery and clothes. On the first payday there was a queue of women wanting to buy electric irons. The recruits were permitted to go to nearby Mosman on Saturday mornings to do any other shopping required and then went back to the barracks for cleaning chores in the afternoon. Sundays were spent going to church for some, sleeping for others, or going to nearby Balmoral Beach for a swim. Some would spend their time writing letters home, washing and ironing uniforms, and spit polishing shoes which took approximately an hour per pair. The women got to know each other and made friendships – some just for the duration of the training and others much longer lasting.

One of the less pleasant things that the recruits had to have was a series of injections – smallpox, cholera and others. This meant that they had to go on needle parade several times where they lined up (just one more time!) outside the RAP and the person at the head of the queue took one step into the room, where a nurse swabbed the recruit's right

arm. The recruit then took another step and was given an injection in that arm, another step forward and the left arm was swabbed, and a further step and an injection was put into the left arm. Recruits would then go straight onto the parade ground for drill and it was quite painful having to hold their arms up to do 'right dress'[2] if they happened to be in the front row.
Lyn Kane Dale

WRAAC School was quite a culture shock for a kid seventeen years old from the bush. I tried so hard but it was very difficult as you had to have everything perfect. Shoes spit polished, giggle dress (the work dress) was linen and it had to be starched and ironed perfectly. Your bed had to be as neat and tight as a drum. They could inspect the beds and drop a coin on them. If the coin didn't bounce you had your bed pulled apart and dumped on the floor. You had to redo it and of course they would check. The drill was long, tedious and tiring. We would march everywhere. You would be at the Regimental Aid Post, getting injections, and then you'd be sent out on the parade ground. The front ranks were always having to put their arms up when they did 'right dress' so you didn't want to be in that rank. When you were in the front rank the arm would hurt so much and just start to drop down and of course you would then be shouted at: 'Get those arms up!' We would have to march to lectures and you had to march past the flagpole which we were told we had to salute.[3]

If you didn't salute it you were in all sorts of trouble. Even if you were on your own and it was in the evening you would look around and see that there wasn't anyone watching but, sure as eggs if you didn't salute you would hear, 'Recruit, get back here'. Where on earth did these spies hang out? I would call reverse charge to my parents, crying and saying I wanted to come home and they would have to withdraw their consent for my being under-age. My mother, a very wise woman, would say, 'Don't worry. As soon as you finish your course and you're posted to your unit you will find it is completely different. You have always bettered yourself with the jobs you've already had'. True to her advice this is what happened.
Betty Hall (enlisted 1963)

Recruit training was fabulous fun and I didn't get yelled at until at least one hour after we had enlisted! I seemed to be a bit of a magnet for such things. Convulsive sobbing wasn't unusual over the six weeks of training as homesickness hit us all, particularly those girls a long way from home, and there was always a long line-up for the phone. Reverse charge calls were the order of the day – anything just to talk to Mum or Dad. My Dad used to reassure me and say 'You'll be OK. This is the best thing you have ever done'. Right again, Dad. Wasn't it such fun learning how to starch our 'giggle' dresses and press them until they could stand to attention unaided? The sense of pride at our 'crisp' appearance, however, soon faded after hours on the parade ground led to chafing and very raw armpits from the stiff and ragged seams of the dresses. Down the hill from WRAAC School was a Citizen Military Forces (CMF) depot and this provided us all with some weekly thrills as we clung to the high chain mail fence and whistled at the raw recruits as they marched up and down the road past our buildings. Life was fun. I really loved marching. Loved it! Being the tallest I was right marker for our March Out and that was fantastic! I nearly burst with pride.
Deborah Harding (enlisted 1969)

The commandos were based close to WRAAC School and would be out early on Sunday mornings running down the road past the barracks to train on the cliffs overlooking Sydney Harbour. We would stand on our verandahs in our baby-doll pyjamas and wave to them as they ran by. They would wave back to us and their sergeant running behind the squad would yell 'Eyes front', although unbeknown to his men he would be running along behind them blowing kisses to us. It livened up our Sunday mornings.
Lyn Kane (enlisted 1964)

We arrived there and met the previous contingent who had arrived the day before us. We were separated into rooms. We were all strangers. This was my first time away from home so I greeted it with some trepidation, but I soon realised that everyone felt the same way and it was an adventure for young women. The training was a great deal of drill on the parade ground. They also tested us for various aptitudes towards different streaming in the Army. I desperately wanted to be a driver. However, because I could type and take shorthand that put me in a clerical position, so my dreams of being an Army driver were never

realised. I found it quite difficult keeping my mouth shut a lot of the time, which got me into a bit of strife occasionally. I found out that it was better to keep my mouth shut than to have to clean latrines at 4 am as punishment. My wardrobe often got thrown on the floor because all the folds weren't facing outwards. It was quite an eye-opener and an experience that really put me in good stead. A lot of those habits that were drilled into me I've never lost, so the training does work.
Sheila Cusack (enlisted 1964)

WRAAC School I found very overwhelming, very different. It was the first time I had been away from home by myself for any length of time and it was just so far from home. Sydney seemed a thousand miles, well it was a thousand miles away – it could have been on another planet. I grew up in a small regional town in South Australia and even though I had been working in Adelaide for several months prior to joining the Army I was still only a couple of hours bus ride away from home. Anyway, I kept my feelings and thoughts to myself. I wasn't going to be a failure. I was going to get through this. However, I soon discovered that if I was to get through the basic training it was a case of interacting with others and putting my hand up when required. The day-to-day chores at WRAAC School didn't bother me, neither did the sitting in the classroom and learning about Army history, policy and procedure. However, the parade ground was a different matter. It just seemed so huge and so much drill – hours and hours of drill - and I had a problem. I just couldn't keep in step. I was always in trouble for square gaiting[4], in fact I think I must have been the worst square gaiter in the history of WRAAC School.
Marion Quick (enlisted 1965)

Recruit training at WRAAC School was certainly different to the life I had at home. I always managed to have a laugh at the things we were required to do, such as hosing the ablutions and cleaning the kitchenette, barracking our bed linen,[5] cheating with the spray starching of our giggle dresses - caught out a few times re-ironing instead of the whole starching process. Learning to march was something else … especially watching my sister square gait.
Andrea Woods (enlisted 1967)

Figure 6 Andrea Woods and other recruits, WRAAC School, 1967.

I remember all the things that everyone else remembers - saluting the flagpole, getting up in the middle of the night to iron our starched giggle dresses because there weren't enough irons, getting our injections and then marching for ages with sore arms to 'pump it through our system', sleeping at night with our arms on a pillow after our smallpox vaccination and walking sideways up steps so as not to crease our uniforms. I received quite a few warnings about the ironing of my dress – they were from World War II and the front plaquet was wrinkled and I couldn't iron it smooth. The seams under my arms were also frayed and the starched

fibres used to irritate my underarms when marching. I ended up bleeding and stuffing hankies into the sides of my bra so as not to stain my dress. I often ended up in tears with the pain but wasn't game to complain. I had long hair and wore it in a French roll but was told that it would have to be in a bun during rookies' training school. I had no idea how to do that, so promptly cut my hair off very short. A couple of my friends also cut theirs the same (we helped each other) and then got into serious trouble for having hair so short that it looked 'manly'. My mother also freaked out! After recruits I was allowed to wear it in a French roll and let it grow with the help of a false hairpiece that I attached and put into a bun until I had enough hair to 'put up'. Don't know why they had different hair rules for rookies.
Kay Owen (enlisted 1964)

My first significant memory of arriving at WRAAC School was when Judy Lucas, from Gawler, SA, said something amusing whilst we were waiting for room allocation. I laughed, only to be severely berated by Sergeant S. when she bellowed 'I have not given you permission to laugh. You only laugh when you are given permission. Is that understood, Private?' I understood! Recruit training was an experience like no other. The bullying, mind games and the attempt by training staff to try and 'break' individuals, psychologically and physically, are all part of ensuring soldiers are prepared and suitable. When you are in the Army, Navy or Air Force you are Government 'property', to do with and go wherever the Government of the day decides.
Maria Adamczak (enlisted 1973)

They can't do to the new recruits what was done in the past, i.e. the yelling and the criticising which you took on the chin anyway because it was part of your training. You didn't take offence at it. Those instructors had to put up barriers between them and us because otherwise you wouldn't learn.
Anna Celi (enlisted 1976)

Two things stand out from my time at WRAAC School. The first was on day two when I got my first defaulters for sticking up for a member of my platoon. I told our corporal he was wrong in accusing one of the other recruits of doing something she had not in fact done. Needless to say I learnt to keep my mouth shut. The second thing was having the

whole mess stand fast because someone went 'Oh there's the Colonel!!' and yes, it was because of my maiden name.
Catherine Sanders (enlisted 1982)

Memories of WRAAC School

Wearing 'giggle dresses' which was our temporary clothing until we were given our tailored uniforms on graduation.

Being on the parade ground and your feet not going where your brain was telling them to go.

Kookaburras laughing. It seemed they would laugh just as you missed a drill direction, like turning left instead of right.

Bed-making lessons and 'how to iron' classes.

Being watched by staff at mealtimes to ensure we were eating properly.

Classes on grooming and behaviour expected of a WRAAC.

Long lines outside the one telephone box from which you could ring out.

Going into Manly to spend our underwear allowance on bras and petticoats.

Learning about the proud history of women in the Army.

Lack of any fitness training. One day we lined up in our physical training gear of shorts and sleeveless tops as well as our white Dunlop volley sports shoes. We ran out onto the oval opposite WRAAC School and had only been there for a very short period of time when an enraged Regimental Sergeant Major (RSM) of 10 Terminal Regiment came storming out on to the oval and insisted we all return to WRAAC School immediately because we were distracting his men 'from their real purpose'. We did indeed return without delay and no further sports sessions were conducted.
Christina Morgan (enlisted 1975)

One day one of the girls said she had seen on the noticeboard that we would be attending lectures on VD. I asked what that was. She said 'venereal disease' and I said I still didn't know what it was. She said, 'You'll find out at the lecture'. Mum, like most of my friends' mothers, had never told any of her children the facts of life. I think she just hoped that we would acquire that knowledge somehow and would save her the embarrassment of having to discuss it with us. However, she would have been devastated if her unmarried daughter had become pregnant as it

was such a disgrace in those days. The day I was packing to go away she came into my room and said 'I just wanted to say to you not to let any of these fellows play around or you can catch things'. I knew she didn't mean that I could get pregnant but supposed that sometimes by having sex you could catch some sort of germ! We went to the lecture and then saw a film which showed revolting images of sexual organs affected by gonorrhoea and syphilis. Finally the advice that Mum had given me before I left made sense. I was not the only one who felt that she was going to be sick and several of the group ran out to the toilets to vomit. I think that film helped many girls to stay celibate for quite a long time afterwards!
Lyn Kane (enlisted 1964)

I adjusted extremely well to Army life. I took to it like a duck takes to water. I had absolutely no issues over discipline and generally did as I was told and behaved myself. I was never in trouble or charged for any offence. I did, however, get confined to barracks for one night as I was caught making fun of a friend who had to clean the canteen windows for committing some misdemeanour. That night saw me cleaning the windows! I enjoyed living in barracks. I enjoyed being able to have as many baths or showers as I wanted - a treat, having come from the country where tank water was scarce, and where the only way to get a hot bath was to light the chip heater. I got on well with the other girls, loved the meals in the mess as the quantity and variety was very good compared with home life. Having said that, I never went hungry at home and Mum was a good, plain cook, but the meals didn't compare with the variety the Army served (especially the sweets).
Marie Taylor (enlisted 1962)

Every morning we four in our room made our beds so a coin would bounce on them. One morning three of us made our beds from scratch and one girl just pulled hers up. When we came back to our quarters after parade our beds were completely stripped except for the girl who had pulled hers up. Grrr!
Gloria Coombe (enlisted 1960)

Denise Dunn enlisted as an officer cadet.

In 1976 it was a six month course run at WRAAC School by women for women so there were no men involved at all. It wasn't the same training that was conducted at the Officer Cadet School at Portsea (Vic.) and it certainly wasn't the same training that was conducted at Duntroon (Canberra) because women weren't allowed to get degrees. In those days women weren't given the same privileges and opportunities. So my six months of training revolved around getting up very early in the morning, cleaning the ablution block, scrubbing floors with toothbrushes, doing all sorts of things. There were about twenty-five of us in the beginning but we finished the course six months later and there were only thirteen of us, so we'd had a few drop-outs. I guess I didn't really learn the sorts of things I thought I would be learning. There was a lot of emphasis on administration and flower arranging – things that I didn't really consider were part of why I wanted to join the Army.
Denise Dunn (enlisted 1976)

Figure 7 Denise Dunn's Passing Out Parade, Officer Cadets, WRAAC School, 1976

Wilma Stralow enlisted in 1955 before WRAAC School was established in Sydney.

We were bussed to WRAAC School, Queenscliff, via Geelong, Victoria and, my gosh, it was cold! Early morning showers in the ablution block some distance away from our hut at 5 am were a real shocker as was the daily routine of drill and classroom lectures. I adapted to the 'life' quite well for a city girl coming from inner Sydney and spent my time spit polishing and learning and complaining of chilblains on my ears owing to the very short haircut!
Wilma Stralow (enlisted 1955)

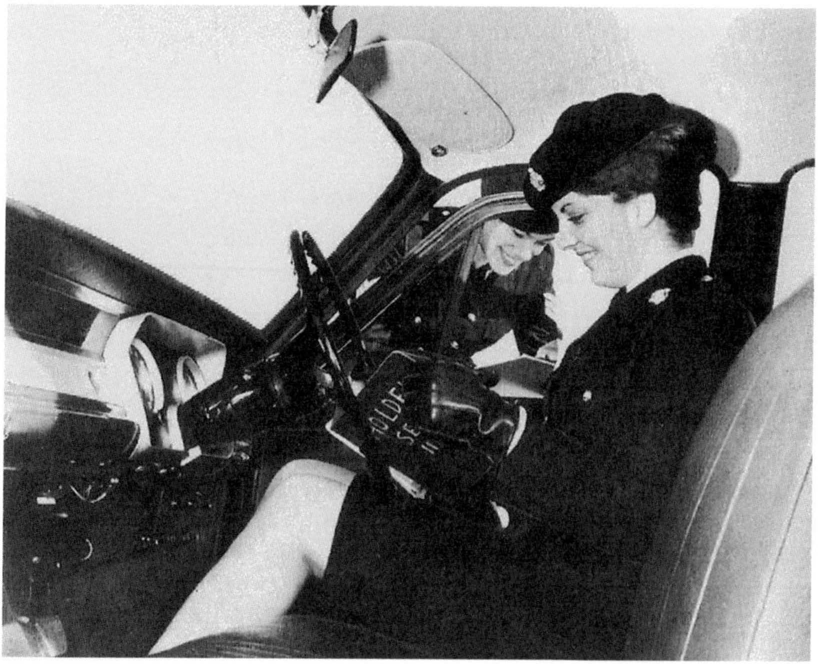

Esma Summerton also enlisted in the 1950s and did her recruit training at Queenscliff, Victoria.

I always wanted to be a driver. I think everybody does or quite a few people do. But of course you get what they decide on after you have a test to see what your capabilities are. My first job was a clerical job which was quite boring so I moved on by putting in for courses and trying to change direction. I became a drill instructor. I was selected through doing a promotion course for sergeants. Back in those days it was all female

training. The recruits did a six weeks course at WRAAC Training Company at Queenscliff which consisted of half a day on the parade ground and half the day in the classroom to learn about rules and regulations. Then they were chosen for a particular trade and they would go on to another unit to train for that. They were very nervous when they arrived like we all were when we did our basic training. The first few days were quite hilarious. They had to learn drill and physical training (PT). I was the one responsible for the physical training at that unit. They learned to march and to slow march, which is a ceremonial march. They have to learn to march so that when they're in the parade they can do it correctly; but it's really a discipline thing, learning to take orders and doing as they're told.
Esma Summerton (enlisted 1955)

Figure 8 Esma Summerton, Physical Training (PT) instructor. Queenscliff, Vic., 1959

Being a Drill Instructor at WRAAC School could be challenging in many ways. Some recruits had difficulty remembering their left and right feet and arms and would send a squad in all directions when they got it wrong. Others would square gait and confuse everyone, but most rose to the challenge of being the best squad and did themselves and the corps proud at their Passing Out Parade. Promotion courses were different. The students were required to learn how to teach drill movements, and to control a squad marching around the parade ground. Occasionally nerves or panic would bring a student instructor undone. On more than one occasion I have watched on as an out-of-control, giggling squad would disappear into one of the gun bays through an open door, or concertina into the back of each other against a closed gun bay door. And, of course, there were those squads who headed off the parade ground and down the grass embankment towards the back gate. All these situations would result in a panicking student instructor yelling 'STOP, STOP, TURN AROUND' or pleading 'COME BACK!!!'

Physical exercise sessions were always part of Recruit and Promotion courses. I learnt very quickly not to include 'star jumps' as one of the exercises the students were required to perform. Unfortunately on promotion courses the student instructors learning to conduct physical exercise sessions were not always aware of the possible effect these 'star jumps' could have on the class (i.e., jumping up and clapping hands above the head as you land with feet apart, then jumping up again, hands down to your side as you land with feet together). This seemingly low fitness activity done continuously for two or three minutes would often result in the participating students producing an audible cacophony of flatulence, and reduce the students and instructor to almost uncontrollable hysterics. The students on one Officer Cadet Course showed initiative when, during drill lessons, they wore under their dark green skirts 'witches britches' (i.e. ladies underwear, like old-fashioned bloomers, with leg lengths down to the knees) with very colourful ribbon, lace or frills sewn around the end of the leg. These garments were not visible during 'quick time' drill lessons, but were very visible and colourful when the squad learnt and practised 'slow time', particularly the slow mark time. They were not part of correct regimental dress, but acceptable on this occasion for modesty, considering the slow, high knee lift required in 'slow mark time'. These anecdotes all created quite a giggle amongst the school staff, instructors and students, and hopefully not too much embarrassment for those involved.
Shirley Ball (enlisted 1965)

Figure 9　Passing Out Parade, 1950s

Figure 10　Parade at WRAAC School, c.1965

I succumbed to pressure and attended my first promotion course. Being married and building a new home were my priority, so ambition and attending full-time training for many weeks was not something I was particularly interested in. I attended the Corporals' Course held at Hampstead Barracks (Adelaide). As there were only two females on the course of ninety-eight trainees, they decided not to place us in the usual accommodation block, opting to hire a very old caravan and put it in the staff car park next to the main building. An order was made that this area of the unit was out of bounds to males, as if that would stop a number of male individuals who seemed to get pleasure out of trying to scare us during all hours of the night by rocking the van. The worst part of this accommodation was night ablutions as we had to use the main building key, turn on the lights and use the toilets. Not practical and definitely not fun. They separated Shelley and me by allocating us different platoons.
Maria Adamczak (enlisted 1973)

I just loved the barracks, the orderly life, the daily routine, loved being tidy and clean, even putting the buttons on the tunic every day.[6] I was so proud on March Out[7] day to receive the 'top recruit' medal. I came first in my intake at WRAAC School.
Sandy Jarrett (enlisted 1980)

Probably the most difficult part for me was learning to get along with so many different personalities. Even though I had been working in the hospitality industry, I was essentially still quite a shy person at this time and had never had to deal with so many individuals in such a small environment before. It all turned out well though, and close friendships and alliances were formed with a few of the other girls while we were at WRAAC School, probably because we had a common 'enemy' in the drill sergeant! I don't have fond memories of the parade ground, mostly because I was extremely uncoordinated and required many hours of extra practice marching up and down (holding on to broomsticks). In the end, I was stuck in the middle of the platoon for our March Out, trying to hide the fact that no matter how hard they tried to teach me, I still couldn't march! I was a 1980 recruit, and as we graduated we were told that we were no longer going to be allocated to WRAAC, but rather to the corps in which we were to serve. I'm sure that our Commanding Officer attempted to instil in us a feel of the 'history' of the WRAAC corps but I think that is something we failed to appreciate at the time. To

most of us, we had just joined 'the Army' as a whole, and we had little to compare ourselves to. There was a lot of talk in the senior ranks about the disbandment of the WRAAC corps, and the fact that we would be the last of the WRAAC to go through the school, but it didn't really hit home until much later what a profound change this would be.
Clare Williams (enlisted 1980)

I took to Army life like a duck to water – loved the drill, hated the needles, enjoyed the classes, the camaraderie, the discipline and the sense of self-worth. Barracks living was okay as I had shared a room at home with my sister but we were five to a room and one person in particular was always slow to do anything. I think she came from a privileged life and expected others to do for her.
Hilary Jones (enlisted 1970)

I adjusted very well to Army life as it was like being at boarding school. I was a day pupil although did spend some time as a boarder in my primary school days. I was used to sharing a room at home as well as in a school dormitory.
Fran Sinclair (enlisted 1972)

Adjustment to Army life frightened the pants off me. I arrived in Sydney to a bellowing sergeant roaring at us about chewing gum. I have never had gum since that day.
Denise Darbey (enlisted 1973)

PT sessions were interesting in Canberra. We were either running around Lake George or we were taken across to the Duntroon obstacle course. With heights not being one of my specialties, I never did manage to get up and over those ropes!
Clare Williams (enlisted 1980)

On the train to Sydney, with other girls, I wondered what I had let myself in for but I quickly settled. I was able to give wonderful reports to family and friends of my new and exciting life! Being a small country-town girl where you knew nearly everyone in town, it was a huge shock to go into the WRAAC. I soon learnt that 'Do as you are told' got me through. Dad did tell me that but I guess I had to learn it for myself. I gained self-esteem and courage to be able to handle the Army way. As I

had been nursing and living in quarters, I found getting used to barracks living came a lot easier for me. I did my six weeks training and stayed at WRAAC School being a stewardess until a course was decided for me.
Daphne Gardner (enlisted 1964)

Figure 11 Liz Anderson in the sports uniform, 1964

I joined the CMF in 1965 and I was seventeen and a half at the time. I started with 14 WRAAC Company at Karrakatta (WA) and after general training I trained as a driver. We used to drive two and a half ton trucks and Land Rovers which was great fun for somebody my size because I was only five foot four. Getting up into those big trucks and driving a

truck that has no automatic gears you had to double shuffle the clutch to change gears. It was a bit of a challenge but we always managed to do it. We used to have some great weekends out in the bush driving the Land Rovers up sand hills north of Perth. On one of the bivouacs[8] we went to Hyden out in the middle of WA to Wave Rock. We used to sleep in the trucks, usually in our gear. Quite often when we woke up in the morning our feet were freezing. They always got us up early but they used to have the fire going and they'd sort out some hot water for us to have a wash, then we'd cook our Army rations over the little stove with the hard fuel fire starter things. Army rations were pretty hard in those days. I can remember those packets of hard biscuits that nearly broke your teeth every time you tried to eat them, and the little can opener and the mushy stuff in the tins. It wasn't the best when we were out on weekend bivouacs but it sustained us so we couldn't complain too much. It was good fun. Our annual camp was on Rottnest Island. I remember the first night we were there and it was dark. All you could see were these beady red eyes down the end of the verandah and the girls squealed because they thought they were rats. When we shone the torch on them we could see that they were quokkas. We'd never seen quokkas before. I went from the CMF to the Regular Army in June 1967. I hadn't been out of Perth before so it was pretty exciting getting on a plane.
Desiree Harms (enlisted 1965)

The first annual camp that I attended I was apprehensive, if not a little scared because I had absolutely no idea of what it would be like. Day one was spent setting up our tents and cleaning up the area. For recruits such as myself this meant the ablution block. The urinals certainly caused some gasps but the toilet cubicles had us all concerned. There were no doors. It was suggested that hessian be strung up to provide privacy for, you know, we are ladies. Then there were our tents. We had metal camp stretchers with a metal locker to hang uniforms and yes, everything had to be neat and tidy. We'd been issued suitcases to carry our belongings and carry them we did, by a handle. There were no wheels in those days. In camp they had to be placed at the ends of our beds and to prevent them getting wet should it rain they would be placed on empty soft drink cans. After the first morning's inspection it was suggested, or that might have been an order, that if all the cans were the same sort it would look better. If I recall we all had to consume a lot of Coca Cola that day. Twelve months after I joined, 1 WRAAC Company went on annual camp. We

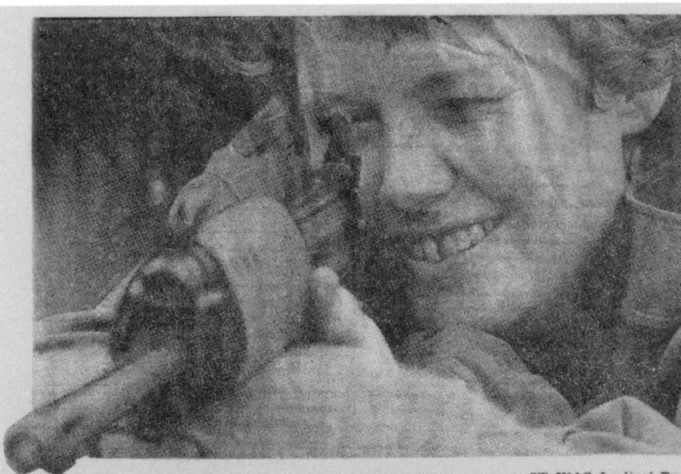

IT WAS Ladies' Day at the Army's Greenbank rifle range yesterday and Private Kathryn Rose, of 1 Women's Royal Australian Army Corps company, received expert advice from Warrant Officer Keith Fall in handling a 7.62 self-loading rifle.

The last of the female Army units in Australia blasted the target boards in the sloppy ill-fitting "comfort" of all male uniforms.

Combat boots, bush hats, webbing belts, baggy trousers and water bottles replaced handbags hats and gloves. Even the shirts the women wore were men's buttons-on-the-right style (picture, left).

The unit will be disbanded on May 31 and its 31 members will be integrated into the male Army Reserve units in Brisbane. Its members began their two weeks' annual camp yesterday.

An Army spokesman said last night that normally a female recruit wore a skirt similar to an air hostess' uniform but this would be impractical in the bush.

The spokesman said the weapon training was to provide an interest "out of the norm" and to ensure firearms safety.

STEPPING OUT *in male uniform at Greenbank yesterday were, left to right, Private Rosemary Guilliatt, Corporal Bethmay Watson and Private Kathryn Rose.*

were invited to volunteer to be trained to fire a rifle. Warrant officers had come from the local training group to instruct us in all aspects of safety of handling and firing a self-loading rifle (SLR). This was a first for women in the Australian Army so the local paper sent out reporters. A couple of days later I was still in camp and my mum opened the paper and here was a photo of her daughter in Army greens marching with a rifle.
Bethmay Watson (enlisted 1975)

I was sent to Mt Martha, Victoria, for twenty-seven weeks to do an Operator Keyboard Radio (OKR) Signal course. It was on this course that I met the soldier who was to become my husband. There were only five girls on this course and fifteen fellows, some married and some in relationships, so I was lucky to find the man for me. After the Signals course the five of us girls were posted to Cabarlah, near Toowoomba, Qld. At that time there were no WRAAC quarters so we were sent to various places in Brisbane and quartered at 10 WRAAC Barracks at Chelmer. In Brisbane I was seconded to Kelvin Grove and Headquarters. It was all so amazing and I was able to use my training a little.
Daphne Gardner (enlisted 1964)

When I arrived at the gates of WRAAC School there were thoughts of 'What have I done'? That was the last time I had time to think of anything except survival. The next two months flew by. I adjusted very well to Army life. It was very organised which suited me just fine. I was good at all the marching, I enjoyed polishing my shoes and early rising was no problem to me. Ironing was not a strong point but practice made perfect. After recruit training my Initial Employment Training (IET) was at Bandiana (Victoria). I had to wait there for over a month until my course commenced. I had to travel to Kapooka (NSW) for a range shoot. I'd never held a gun in my life let alone fired one. We were told over and over at WRAAC School that WRAAC were ladies. I think they forgot to tell the rest of the military. I was the only female doing the Basic Storeman course and I finished it with little effort. I wonder how, when I think back, as I really didn't have a clue what the instructors were talking about. They could have been talking Martian for all I knew. It was during the course at 1st Training Group that my time as a WRAAC ended. I was happy with the pay rise but now I had to do my corporals' course with the men. It was at Canungra, Qld and there were eighty males and two

females. I was lucky to be fit enough to keep up with the men at PT. I was also lucky to have a huge voice so shouting was no problem either. My second and final posting was to 2 Army Recruiting Unit, Sydney. Although I was no longer a WRAAC I like to think that I upheld the values of always being a lady in uniform. My military career ended on 23 January 1986.
Annette Wilson (enlisted 1977)

I did a clerical course at Broadmeadows (Vic.) and with those skills went to Puckapunyal. It was the time of the National Service and I helped with the eye testing, payroll and other clerical duties. We were the first girls to participate in the twenty-mile route march carrying packs and rifles the same as the men. It was really tough, but we all completed it and were very proud.
Ann Dullaart (enlisted 1966)

Passing out day April 1964

Figure 12 90/64 Passing Out Day, WRAAC School, Sydney, April 1964

Much of the recruits' time was spent preparing for their Passing Out Day. All the hours of drilling on the parade ground and the time spent washing, starching and ironing uniforms culminated in this day where everyone was perfectly turned out with beautifully starched new summer uniforms, spit polished shoes, straight stocking seams and not a hair out of place under our berets. We marched proudly onto the parade ground in front of all the staff of WRAAC School, the visiting Army brass, as well as family and friends. We were put through our paces, all perfectly in time to the band music, before being inspected by a visiting male Army officer of suitable rank. We had been told that if an inspecting officer addressed an individual she should look straight ahead, reply to the question and always remember to add 'Sir' or 'Ma'am'. On this occasion the inspecting officer halted in front of one of the women and asked her a question which she answered and added 'Ma'am' instead of 'Sir'. As she said later she had never spoken to a male officer before and completely forgot what she was supposed to call him. After the parade was over and the visitors had gone we were told where we were to be posted. Not everyone passed the course. We had begun with thirty-one girls and at the end there were only twenty-five. That afternoon we packed our cases. Most girls left that night to go to their new units but six of us were not leaving until the following day. We were allowed to go in to Central Station in Sydney in the truck to say farewell to the girls who were leaving. I vividly remember us all in uniform on the station hugging and kissing each other, crying and waving as the train left. It was sad to say goodbye to friends who had become like family in those few short weeks and I felt quite lost and alone afterwards. Those of us remaining had to spend the night in our quarters which were suddenly very quiet.
Lyn Kane (enlisted 1964)

At the end of the training when we were told that we had passed I remember that I was just so excited I rang my sister in Monto (Qld) and told her that I'd just been promoted from recruit to private and that I was marching out to 3 Camp Hospital, Puckapunyal in Victoria.
Ruth Grace (enlisted 1964)

Figure 13 Passing Out Parade, WRAAC School Sydney, 1964. Cherryl Burrows second from the left, middle row.

the UNIFORMS *you'll wear with pride*

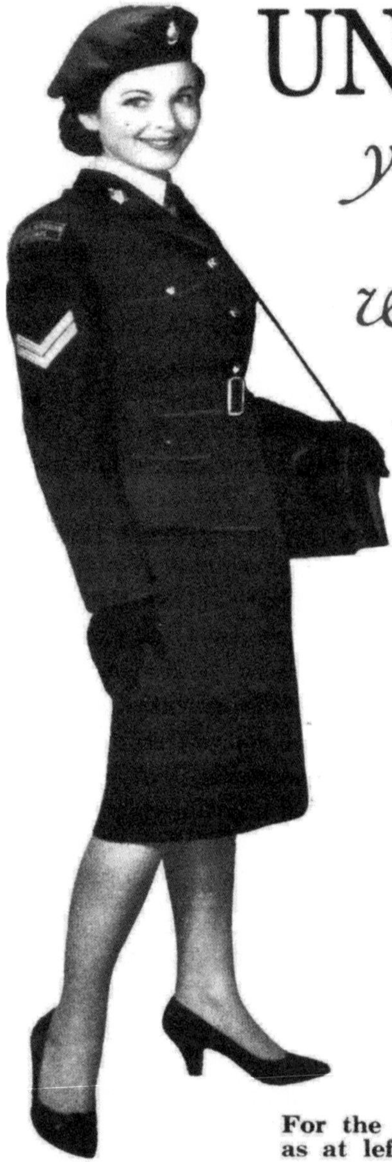

For the occasion, whether walking out, as at left, or driving an Army vehicle.

2B Uniforms

Here is some of the range of summer and winter outfits, of excellent design and quality, fitted for you, and free. There is an allowance of £18 also for Regulars to provide their night attire and underwear. The winter uniform is an attractive Highland Green twill suit and comprises jacket and skirt with gilt buttons worn with shirt and tie, and beret with gilt badge pinned over the left eye. The summer dress is parchment creaseless linen, worn with felt hat and green tie or open neck.

ITEMS OF ISSUED DRESS

2 winter uniforms
1 greatcoat
3 shirts
1 fawn felt hat
1 cardigan
1 pair gloves
3 ties
5 pairs stockings
3 pairs shoes
3 summer frocks
1 beret
1 mirror
3 towels
5 brushes
1 pair sports shorts
1 sports blouse
2 pairs ankle socks
1 pair gym. shoes
6 handkerchiefs
1 attache case
1 suit case
1 shoulder bag

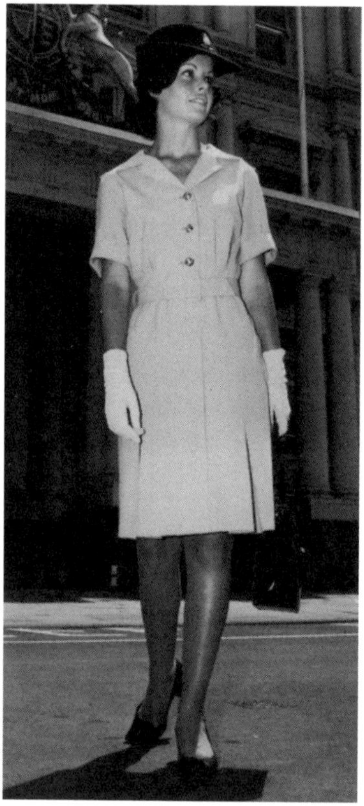

Figure 14 (left) Mary-Anne Larmer in WRAAC summer uniform, c. 1969. Army publicity photograph.

Figure 15 (right) Mary-Anne Larmer in WRAAC winter uniform, c. 1969. Army publicity photograph.

We all went from WRAAC School, the whole lot of the recruit course that I was on, to the city to get measured up for our uniforms. In those days most Aussies didn't have extended wardrobes like they do now, so we all donned our flashiest city clothing. For me this was a mustard corduroy suit with fake fur cuffs that I'd worn on the flight over from Perth to Sydney, my one pair of ultra-high heel shoes, 15 denier nylon stockings with the seam up the back and worn with suspenders. Pantyhose hadn't been invented then. My hair was done up in a lovely

French roll. The mini skirt phenomenon hadn't hit Australia at that stage so our skirts still came to a few inches below the knees and they were quite slim fitting. We all clambered up the ladder on to the back of a big Army truck for a ride to the tailor's and it was quite a trick in those flash city clothes. I clearly remember the day a few weeks later when we were all issued with our brand new uniforms. How incredibly exciting it was. I did feel proud to wear the uniform, in particular the winter uniform, which was a dark green with gold buttons and gold WRAAC badges worn with a green lanyard, mushroom coloured shirt, dark green tie and a green beret. Very, very flash.
Helen McLeod (enlisted 1963)

They didn't wear marching out uniform during the day while they were training. They wore giggle dresses as we called them which were a khaki cotton, very hard to keep clean and very hard to keep starched and pressed and they had to put a lot of work into the dresses to make them look decent.
Esma Summerton (enlisted 1955)

In the mid-1960s pantyhose came into being and it was such a joy not to wear suspenders and seamed nylon stockings. Kayser stockings were quite expensive, so I started to wear pantyhose and drew a line down the back of my leg with an eyebrow pencil. No-one seemed to notice and we often just didn't worry about the line. One day on parade at the Supply Depot, the Assistant Director, WRAAC, pulled a surprise inspection and immediately noticed we had pantyhose on. She asked our boss (male Major) why we hadn't been pulled up over this issue and he commented that he 'didn't like to look at the women's legs too closely'. There was much snickering in the ranks (male mostly) and we were given a good talking to! If one leg of the pantyhose got a ladder in it, we simply cut it off below the 'panty' part and wore two sets with one leg each and saved money!
Kay Owen (enlisted 1964)

During trade training we were accommodated at 10 WRAAC Barracks at Enoggera, Qld. There was a dedicated date across Australia for soldiers to change either into summer or winter uniform. As trainees, we had not yet received our winter uniforms and therefore had to wear men's battledress trousers and 'bomber' jackets when the changeover

into winter uniform came in. These did not look the best on us as the uniforms had been made for men with a large girth and we were all slim. I would receive my WRAAC winter uniform from the Area Clothing Store at my first unit, tailor-made just like the summer uniform.
Christina Morgan (enlisted 1975)

Figure 16 (left) Ruth Grace in RAANC walking-out winter uniform, c. 1965. Army publicity photograph.

Figure 17 (right) Ruth Grace in RAANC ward uniform, c. 1965. Army publicity photograph.

Bendigo was the only place I ever had to do a 'great coat' parade! I can recall trying to adjust buttons on one of the girls' coats as she had increased quite a bit in the bust and the buttons would only just fit on the edge of her coat. After the parade she did a swap with another girl who had lost weight.
Hilary Jones (enlisted 1970)

It was a really nice uniform. We had a summer uniform and the hem had to be a certain length from the floor when you had your shoes on. Nice black shoes that we had to learn to polish and lisle stockings[9]. Goodness knows what girls would wear today. I don't think they would wear this sort of stuff. But the summer uniform was a funny greenie sort of a colour and then we had a nice winter uniform which was a skirt and a long top. We had a greatcoat and we had a beret to wear with our winter uniform and a hat with our summer uniform. When the girls were all dressed up we looked really nice.
Margaret Elder (enlisted 1953)

I remember feeling really good about myself whenever I ventured into the world in uniform. I loved the way people would glance my way and how their glances would linger. I was young, slim and pretty and I wore my uniform with pride.
Marie Taylor (enlisted 1962)

In the 1980s the female uniform changed from the khaki we had been wearing to a lime green dress. It was quite a change, both good and bad. The good was that we could buy something off the rack and I soon had six uniforms so that I could do a mass iron once a week. The bad thing about the new uniforms was that I felt quite disassociated from Defence as the uniform did not seem very military in its design or colour.
Christina Morgan (enlisted 1975)

WRAAC's do many interesting jobs

- **Clerk**
- **Draftswoman**
- **Signals operator (including cipher)**
- **Psychological coder**
- **Cook**
- **Storewoman**
- **Driver**
- **Stewardess**
- **General dutywoman**

The WRAAC trains you in all these vocations, on full pay, from joining. You have to show character, intelligence, and enthusiasm, for these qualities are the backbone of our Army.

WIDE CHOICE OF CAREERS

Postings: During your initial training you will be selected for a posting to an Army unit which has a vacancy most suited to your particular ability and choice. If you have a preference to serve in a particular State or area, this too is considered when postings are planned.

The WRAAC offers you a wide choice of interesting occupations. Training for these is conducted at specialist courses at Army Schools or by "on the job" training in your unit. Included among the WRAAC employments are:—

Driver
Projectionist
A variety of clerical duties
Cook
Psychological Coder
Psychological Examiner
Draughtswoman Topographical
Draughtswoman Lithographical
Operator Keyboard
Operator Keyboard and Radio
Operator Keyboard and Cipher
Operator Signals

Chapter 3
The Postings

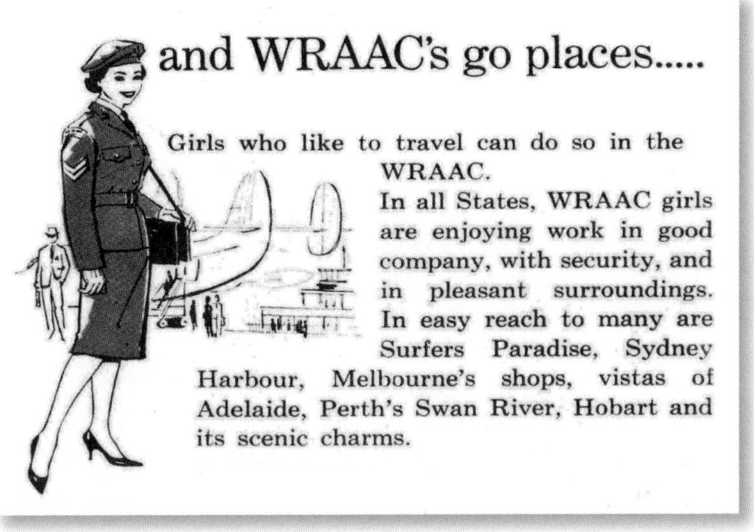

After recruit training the women dispersed to their postings all over Australia. Sometimes they met up again but for most their paths never crossed again. With postings being for around eighteen months it would usually be guaranteed, however, that when someone marched into the barracks near their new unit, they would find a familiar face. It was always great to see a familiar face. From WRAAC School some went straight to their new units while waiting to go on trade training courses, but many others were sent to do Mess Stewardessing/General Duties (GDs) during the interim.

The Army in its wisdom posted two people with the same surname to the same unit (7 Camp Hospital, Kapooka, near Wagga Wagga, NSW) - a nurse named Christine Cain and me. From memory it took Chris and me about six hours to travel by train from Sydney. The driver who picked us up at Wagga Wagga railway station smirked as he asked 'Cain & Kane?' We were known as Private Cain with a C (for Chris) and Private Kane with an L (for Lyn). It was inevitable that the time came when someone asked me 'How do you spell Kane with an L?' I was to do General Duties and Mess Stewardessing while waiting to go on a Clerk Ordnance course. I was sure I wouldn't have to wait long until I went on to my 'real' posting. Little did I know I was to spend the next eleven months there. I knew that my next posting would be to an ordnance depot in one of the capital cities. In the meantime I was happy to fill in the interim period, usually about eight to ten weeks, serving meals in the mess, sweeping and washing floors and doing general cleaning. It was boring, monotonous shiftwork. I tried to be positive but I grew to hate the job and felt very isolated as Kapooka, the training camp for all male recruits, was situated in the bush seven miles out of Wagga Wagga. There were approximately 3,000 men and thirty-three women in the camp. The male recruits were quartered at 1 Recruit Training Battalion about a mile away from the female quarters which were next to 7 Camp Hospital. The women were not permitted to walk unaccompanied around the camp, especially at night, and we had to get the duty vehicle to take us to the main canteen or to the movies at 1 RTB.

I was a city girl and felt like a fish out of water in that environment. It was very daunting to walk with several other girls into the main canteen at 1 Recruit Training Battalion for the first time. Several hundred recruits were in there and many stopped talking and turned around to watch us as we walked in the door. I was so self-conscious I thought I was going to trip and fall flat on my face. It got easier after that and by the time I left Kapooka I could walk into the canteen and hear my name called and guys whistling but would just be able to ignore it. No-one bothered us. I guess like most young men they just wanted to see some girls as they had to stay in the camp for their first six weeks of training. Female personnel were strictly forbidden to go out with recruits (bad for discipline) or the young officers (for the same reason), though of course many did. We were permitted to go out with sergeants and other ranks (corporals and privates). Many of these men had wives and families elsewhere, which they sometimes conveniently forgot to mention. Although I was surrounded by thousands of men, the time that I spent at Kapooka was the loneliest year of my life. I worked shiftwork and had my days off during the week. On weekends when most of the girls were at the barracks relaxing and having fun I was at work so didn't get a chance to mix with many people. I couldn't wait to get out of the place.
Lyn Kane Dale

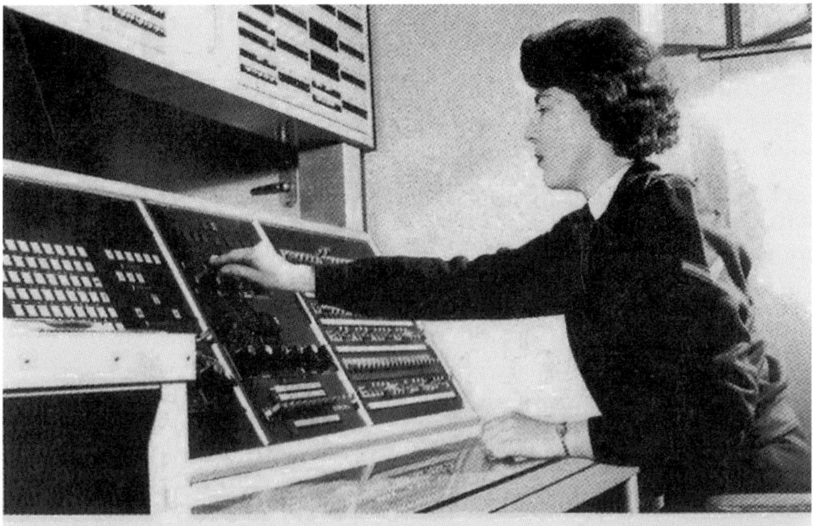

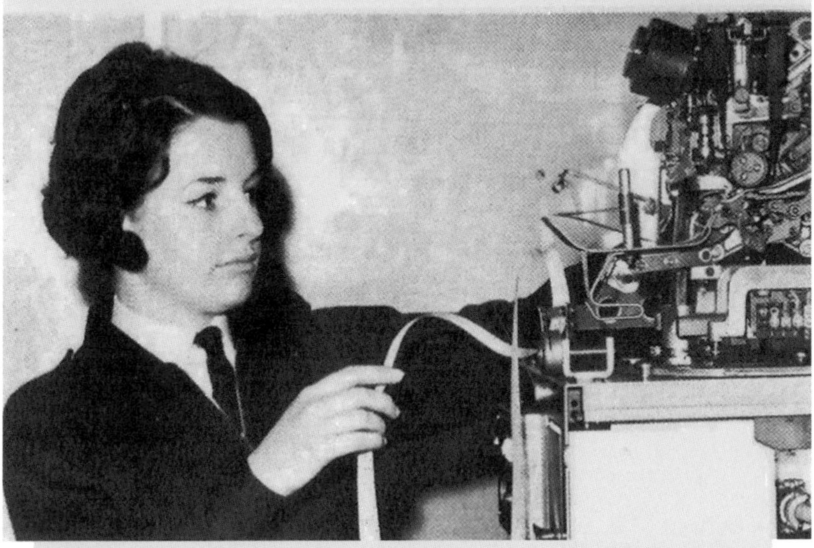

MEDICAL CARE

The service ensures that the finest medical, dental and optical treatment, including the very best in specialist attention and hospitalization is available for all its members. The cost of all medical care is borne by the service—there is no expense to the member, who continues to receive full pay during the period of sickness and convalescence.

Kapooka was a great place. It was cold in the winter and stinking hot in the summer, and in those days the offices were old munitions huts so no air-conditioning. But anyway, it didn't matter because even though you were working it was still great fun. I was the Movements Clerk which meant I was responsible for booking and all the paperwork involved in sending the recruits, once they had marched out, from one end of the country to the other.
Jenny McLay (enlisted 1962)

Firstly they made a mistake. After sending me to 1RTB, Kapooka, they put me in the pay office and found that while I was quite bright in some areas mathematics wasn't one of them. I just don't have a brain that works very well with numbers. But I found my niche. They sent me over to be an assistant to Major Kelly in Training Wing which was just the right place for me because I found that the work that I had to do there was exactly what I was really very good at. So they did finally put a round peg into a round hole. While there were still parades and room inspections we had more freedom than we had at WRAAC School. We still did have closed camp nights and were bussed into church halls on Sunday night to go to church dances and things like that, which we found quite hilarious at the time. However, I imagine that was their way of keeping young women protected who were in their care.
Sheila Cusack (enlisted 1964)

I served in the RAANC at 7 Camp Hospital 1RTB, Kapooka, in 1964-65. As this was the Recruit Training Battalion there were always a variety of ailments including upper respiratory tract infection (URTI), so penicillin jabs and inhalations were the usual treatments. I think the penicillin must have been leftover from World War II as we had to mix it with sterile water to use it. One occasion that sticks in my memory was when a non-commissioned officer (NCO) came in for treatment with a rather nasty infection after he had undergone a circumcision. Being a rather naive 18-year-old and having never done anything like this before I didn't know where to look and just wished the floor would swallow me.

Naturally I wanted to look as if I knew what I was doing as I prepared the sterile dressing and I managed to apply it with some professional diplomacy and hoped that I wouldn't be on duty for his next treatment. **Chris Cain** (enlisted 1964)

Figure 18 Sheila Cusack, 1 Recruit Training Battalion, Kapooka, NSW, 1964

My first posting was at WRAAC School as a GD or mess stewardess. I was quite furious at losing a good telephonist job to do cleaning and toilets and things like that. I went to the CO and basically demanded that I wasn't going to do this job. I didn't join the Army to do general duties. I got told to do as I was told. I was at WRAAC School for probably eight months and I thought it was never-ending and I was off and on complaining. The School Sergeant Major (SSM) came to me one day and asked if would I like to do a basic signals course so from there I went to School of Signals, Balcombe (Vic.). We learnt to type on the basic teleprinters and they were the old fashioned TG7B models with a box over the keyboard so that you couldn't cheat and you had to learn to touch-type. From there it was a cipher course and then I ended up as a supervisor at Victoria Barracks, Sydney at Paddington.
Julie Elverd (enlisted 1960)

I travelled by train with some other girls to Melbourne for my first posting to Central Army Records Office (CARO) at Albert Park Barracks. I was to live at 31 WRAAC Barracks in St Kilda Road and it was all very exciting and different. I felt quite lonely for a while as I walked to work alone and lived in a room with other girls who weren't very friendly. Weekends I spent with other girls from rookies and we enjoyed that time together. I was glad not to have been posted as a GD (General Duties) but seemed to work with mostly civilians at CARO.
Kay Owen (enlisted 1964)

After recruit school, I was posted to HQ 1 Division in Paddington, behind Victoria Barracks, Sydney. I was there doing clerical work for three months until being sent to Puckapunyal on a clerical course for two months. After Pucka, I was posted to 402 Signal Regiment in Dundas (Sydney), and that's where I worked until my discharge, so there were no more postings. During the term at 402 Sig Regt I completed a Junior NCO Course and was promoted to corporal. I later turned down a Senior NCO Course and subsequent promotion to sergeant as I was about to be married. Had I not married, then I feel sure that I would have re-enlisted for another three years. I was never posted overseas. Those postings were as rare as hen's teeth back then. I didn't know anyone who went overseas.
Marie Taylor (enlisted 1962)

Figure 19 'Clowning around'. Julie Elverd in mess stewardess uniform with serviette on her head, Officers' Mess, WRAAC School, 1960

PROMOTION OPPORTUNITIES

Opportunities for promotion in the Corps are excellent. Each of the WRAAC trades carries higher rates of skilled pay to which you can progress as you gain higher qualifications and experience. Promotion to the non-commissioned officer ranks, is open to any girl who has the desire and ability and you receive every assistance and encouragement. Many WRAAC officers are promoted from the ranks of the Corps. The Officer Training course commences in April each year. This 27 weeks course is open to members of the WRAAC and civilians who are between 19½ and 30 and have an education standard of not less than Junior or Intermediate Certificate.

RATES OF PAY

Full details of the rates of pay and allowance for members of the WRAAC and RAANC are given in a separate publication— "Pay Rates in the Women's Services—Australian Regular Army", available at any of the Army Recruiting Centres. In addition this leaflet explains the pensions and gratuities provided at the completion of WRAAC service.

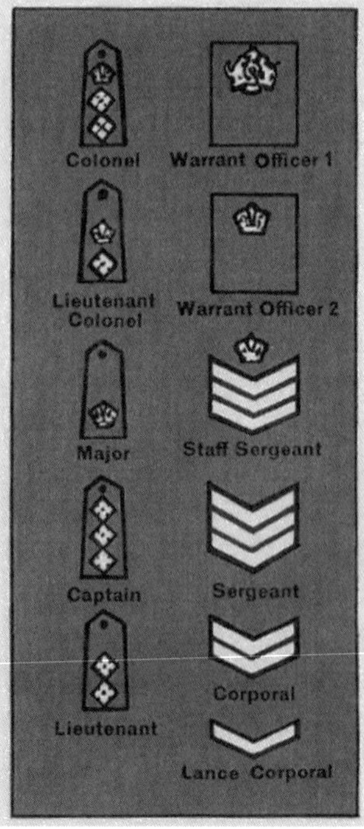

Most of my Army work was clerical with a bit of driving thrown in. My first posting was 21 Cadet Battalion (living at 31 WRAAC Barracks) for about nine months and doing a camp at Puckapunyal during that time. I then went to 3 Military Police before heading off to do a clerical course at 30 WRAAC Barracks at Mt Martha, Victoria, from July to August 1963. I was then posted to 2 BOD Moorebank, NSW, where I stayed until my discharge in 1965.
Julie Ross (enlisted 1962)

My first posting turned out to be a Recruit Instructor. It was really wonderful. I stayed with Headquarters of the WRAAC Company until I got my commission and then went to Signals. My father was very thrilled. In the Regular Army things were different. When I was posted to Central Army Records in Melbourne, the warrant officers were very anti-WRAAC (WRAAC officers anyway). After a couple of months they accepted me and got used to me, after seeing that they were not going to wear me down. It was at a dinner in the officers' mess at Eastern Command in Sydney where I met Bill before he went off to Vietnam. Later I was posted to Melbourne and attended a lunch at Victoria Barracks officers' mess, when a gentleman (Bill) walked up to me and said, 'I know you'. Three years later I married him.

My husband was posted to Adelaide as Chief Engineer. We loved Adelaide. My first job there was for two weeks initially to stand in for the Camp Commandant at Keswick Barracks, who was going on leave and then I was to take over from Cath Cruikshank at Command HQ. She was retiring and still had a month to go. A week before the Camp Commandant was to return from leave, he had a coronary and was off work for a further three months. I stayed in the position having to be formally 'posted' to the appointment for legal requirements because the role of Camp Commandant was to initiate Courts Martial. I was in the position for three and a half months. The *Adelaide Advertiser* interviewed me and an Adelaide radio station asked if I could come in to be interviewed as 'Woman of the Week' - I was a bit of an oddity, as no woman had been a Camp Commandant before. It was a wonderful job with wonderful staff and I enjoyed it tremendously. I had to organise the pay for everyone, not only at Keswick Barracks but for outside units as well. One requirement was that I was supposed to carry a revolver and of course women weren't allowed to touch firearms, so the sergeant major carried the gun for me and accompanied me to do the pay. Cath Cruikshank eventually retired and I took over her job.

While I was at Keswick Barracks I was appointed to a number of Courts Martial Tribunals – very interesting actually. This was just after our troops were being withdrawn from Vietnam so we had a number of young gentlemen, National Servicemen in particular, wanted for desertion or extended Absence Without Leave. It was very interesting to be on a tribunal and to have the responsibility of determining what the soldiers' fate would be. I think we discharged most of them, however, we sent one of them for ninety days to the military prison in Sydney though that was for more than desertion. It was definitely an eye-opener to see how military justice really worked. The courts were run very well and were very fair. The person representing the Army, the prosecutor, was not a trained lawyer. He would just be a normal officer though he normally worked in the legal support area, while the person charged could have a barrister working for him to support his case. There were three officers on the tribunal and the junior officer always had to give their verdict first and the senior officer at the end so there would be no coercion of the junior officer by the more senior ones.
Iris Petrass (enlisted 1961)

I thought when I pulled in on the train at Brisbane it was like a big country town. They talked a different lingo. The chap said to me 'Where is your port?' and I'm looking around. Well, I'm at a railway station not a shipping dockyard. He must have seen the confusion on my face and he said 'Where's your case?' I did two and a half years in Brisbane and thoroughly enjoyed it up there.
Christine Negus (enlisted 1963)

I was lucky enough to graduate top of my class and I literally picked up my bags from the cadets' lines and walked down the hill to the officers' mess. I was posted immediately back to WRAAC School where I became a Recruit Instructor, a recruit Platoon Commander and I also provided some lessons to the officer cadets. So, here was I all of six months in the Army, about to be an instructor and knowing absolutely nothing about the Army outside the gates of WRAAC School. It was a little bit daunting and I think I was the last one that actually happened to because I remember speaking to the commanding officer at the end of my first year

as an officer, saying 'It doesn't really inspire confidence in me or in any of the recruits or cadets who were going through because I have absolutely nothing to offer them about what it's like on the outside'.
Denise Dunn (enlisted 1976)

Before our Passing Out Parade, we were given our job allocations. Six of us were told we were going to Canberra as Electronic Data Processing Officers. I had no idea what that entailed. I trained as an Operator Keyboard and then Cipher. We were billeted in Gowrie Hotel, Canberra, commonly called the 'Twin Towers of Sin', due to accommodating all female soldiers (there was no WRAAC barracks at Duntroon), ACT police, Australian Federal Police and public servants. It was like living in a palace. I remember once getting into the lift in uniform and saluting a Warrant Officer Class 1. She was the Regimental Sergeant Major (RSM) of the corps at that time and she politely said, 'Thank you Private. I appreciate your respect, but I am only a Warrant Officer'. From then on every time I saw her we always acknowledged each other with a smile and a nod.
Denise Darbey (enlisted 1973)

I clearly remember being advised of my posting. 'Private Harding, you will be happy to know you have been posted as a Mess Stewardess to 2 RTB (Recruit Training Battalion) Officers' Mess, Puckapunyal, where the ratio of males to females is one hundred to one.' She was right, I was happy, but not necessarily because of that ratio. Real adventure was to begin. Life at Pucka was an absolute blast!
Deborah Harding (enlisted 1969)

Someone, somewhere deep in the system decided that I would be well suited to clerical work. Clearly they were either deeply misguided or had a very ironic sense of humour. My first posting was to Apprentices' School, Balcombe (Vic.). Aside from being told that the CO didn't want 'skirted soldiers' on post it was a good time. Looking back I see that he had responsibility for hundreds of healthy (OK - oversexed) young men, so the presence of a number of young girls must have filled him with horror about what might eventuate. I felt sorry for many of the apprentices – they were only kids themselves and they were quite often bullied by the training staff, many of whom were, I now realise, veterans of World War II or Korea and I think they probably had lots of personal issues going on

which can't have helped the situation. I travelled to Apprentices' School by train. I can still remember how many people stared at me (maybe a female in uniform was a novelty for civilians). It was freezing cold and I was glad to have that huge, heavy dark green greatcoat to sleep under.
Kim Lake (enlisted 1966)

Although I did the clerical testing along with the rest I managed to get selected for cartography. I believe I was lucky. Three others from the same intake also went to Bendigo with me. My posting to Bendigo, Victoria, and the Survey Regiment was an eye-opener for me. Because there were four of us one did GDs, one went on course straight away and two did on-the-job training. The three remaining in Bendigo swapped each week until it was discovered if I continued cartography I would be blind within three years – we worked so close to the mapping sheets while scribing over a light table – so I had to change my job or I could resign. I chose General Duties draughting – once again I believed I was very lucky. In November of 1970 I started corps training at the School of Survey which was situated at Bonegilla (near Albury) on the Hume weir. In February 1971 I was posted to 6 Signal Regiment, Watsonia (Melbourne) as the junior draughtswoman. My first task really had me panicking as I had to paint a mural (on canvas) as a backdrop for a function in the Sergeants' Mess! This was to be a beach scene on a canvas thirteen feet high by twenty-five long! Move over Rolf Harris! After the success of this I was in demand for any posters, paintings, etc. that were required for any reason along with all the mundane work of wiring diagrams, signs, schematics, floor plans, etc. At one stage I went to work at 127 Signals Squadron at Ivanhoe and I was the only female on site. Quite a learning curve that was. At that stage I was a corporal and when I returned to 6 Sig I was in charge of the drafting office.
Hilary Jones (enlisted 1970)

About three months after I arrived at Kapooka some of the nurses went off on courses and so did one or two of the GD girls who had come there after me but I was still there doing General Duties. I thought there must have been a lot of girls waiting on Clerk Ordnance courses and that it would be my turn soon. Another couple of months went by and more girls came and went and so I requested an interview with the WRAAC officer and asked her why I hadn't been sent on a course when others who had come after me had gone. She told me it was not up to me to question what the Army was doing and just to obey orders. The Army knew best and could give me any job that was deemed appropriate and I needed to prove to her that I could do this job well. She told me that no Clerk Ordnance courses were being run at the moment. Several more months went by, several more girls came to do GDs and left after a few weeks and I was still there. I again requested to see Lt C. and got the same story about proving that I could obey orders and was told to be happy with what I was doing. I told her I hadn't joined the Army to wash dishes and sweep floors and she became angry and told me I had no choice but to obey orders and get on with it. I knew it hadn't made the situation better by challenging her but I was really frustrated and very unhappy. I couldn't go over her head as all contact with the Commanding Officer had to be through her.

Finally ten and a half months after I arrived at Kapooka I made the decision to go Absent Without Leave (AWOL). I wrote to my parents in Perth telling them that I was taking some leave, going to Sydney and would not be returning to Kapooka. I said that I would be in touch again after I had found a job in Sydney. Several days later my room mate came into the mess for lunch and quietly told me to come up to our room as she had something to tell me. When we were alone she said that Lt C. had been called to the CO's office during the morning and the nearby personnel could hear him shouting at her and that my name had been mentioned. Several minutes later she came out of his office and ran crying to her office. My room mate told me she would let me know if she found out anything else and later that day told me a signal had come through from Canberra confirming that I had been posted to 3 Base Ordnance Depot in Melbourne. I started packing straight away. We didn't know what had happened until several days later when I received a letter from my mother who told me that when she got my letter she contacted Capt Gardner, the WRAAC officer who had interviewed me when I applied to join up in Perth. Mum went in to Swan Barracks to see her and told her that as she (Mum) had signed the permission form for me to enlist she now wished either to withdraw her permission because I was under twenty-one years of age or for the Army to post me immediately into the position I had been told I would be allocated to. When I arrived at my new posting I discovered that there were no Clerk Ordnance courses. Everyone learned on the job. Mum had saved the day!
Lyn Kane (enlisted 1964)

To get schooling, to be well fed and be posted to a clerical position at headquarters Southern Command in beautiful old Victoria Barracks, Melbourne, not to mention the sheer thrill of living at 31 WRAAC Barracks and walking to work along misty, beautiful St Kilda Road each day or hopping on a delightfully rattly old tram for the short trip up the road. I was in seventh heaven. Army life was a breeze for me, coming from a very strict home life. I loved the freedoms and the privileges it gave me not to mention the security, both financial and otherwise. I loved it.
Helen McLeod (enlisted 1963)

I was sent to 3 Camp Hospital, Puckapunyal, and I spent the rest of my time there. I did my nursing training at the School of Army Health in Healesville (Vic.) and gained my Cert II. I also saw my first snow on a day at Mt Buller and found out what being really cold was like. On returning to 3 Camp Hospital I was promoted to corporal and spent the remainder of my time as the senior nurse in Casualty. I very much enjoyed the friendship and camaraderie of the other nurses, both male and female. My least favourite shifts were the 'nights' when the hospital was silent and every sound scary. The morning seemed to take forever to come.
Elizabeth Anderson (enlisted 1964)

Six of us went off to Bandiana (Vic.) for our admin course. The contrast in the weather was unbelievable. One minute we were on parade in the fog, being yelled at for shivering and the next minute we were boiling. After Bandiana I got posted back to HQ Training Command as the CO's clerk, but ended up on the signal desk. Working in Darlinghurst, around the corner from Oxford Street was very enlightening. I moved with Training Command to what was no longer WRAAC School. This was a very different experience from when I was originally at Middle Head. We were given permission to park on the parade ground but this is something I could never bring myself to do. It went against everything we had been taught. My last memory of Middle Head was sitting on the bar with a line-up of rum and coke.
Catherine Sanders (enlisted 1982)

I was posted to the School of Military Engineering at Casula (NSW) at the beginning of 1977. It was a traditional posting time and I was part

of a number of new staff. Those who were not Royal Australian Engineer Corps soldiers were put through a local induction to learn what the Engineers did. Over a three-day period we operated plant equipment, went down the Georges River on river craft, learnt about bridging, saw how the dogs detected explosives, etc. It was a great introduction to the unit and this type of induction was well before its time.
Christina Morgan (enlisted 1975)

I was in the Army for seven and a half years, at 2 Base Ordnance Depot Moorebank (NSW) the whole time, but rotated within the depot with different jobs. I started working in the storehouse, moved to clerical in the control office and then as a Depot Clerk within the area. This job allowed me some driving duties. Finally made Corporal about 1966 as promotion was difficult due to the small number of WRAAC at the depot. We were always treated very well by the men at the depot. There were still some World War II and Korean War Warrant Officers and sergeants there.
Merle Viles (enlisted 1962)

I had requested to be an Army driver and was given a commitment by the recruiter that this request would be honoured. Little did I appreciate that, once I signed on the dotted line, they owned me. After recruit training I was allocated a Clerical Administrative position. I definitely did not want to continue working in an office but I did not question this allocation. I unfortunately rarely questioned anything in my younger years.
Maria Adamczak (enlisted 1973)

Those of us who were joining Signals were transported by Hercules aircraft from Sydney to Melbourne to start life in the regular Army at Watsonia Barracks. Becoming a Signalwoman was a bittersweet experience – while it felt on one hand that we were becoming equal with the men, it also felt like we were losing some sense of history. With the loss of the separate WRAAC corps I don't think a lot of us appreciated what a huge change it was for existing WRAAC. Getting to know people this time was a bit more difficult, as we were faced with those who had been in the Services for some time and who gave the 'rookies' a hard time. I found it much more difficult to 'fit in' here, but if all else failed at the end of a hard day, there was always the option of joining the rest of

the guys up at the boozer to watch yet another episode of MASH.
Clare Williams (enlisted 1980)

Because of my clerical background I was selected as a clerk. I was so upset about that as my whole life I had been told what to do. The day I was advised I went straight to the head instructor whose name fails me although I can remember her face very clearly. I told her I wanted to go home as I could be a clerk back there and that I joined the Army to be a driver. I did get to be a driver and waited at WRAAC School as stewardess until our driver's course began in October of 1972. After driver training I was sent to Kapooka.
Fran Sinclair (enlisted 1972)

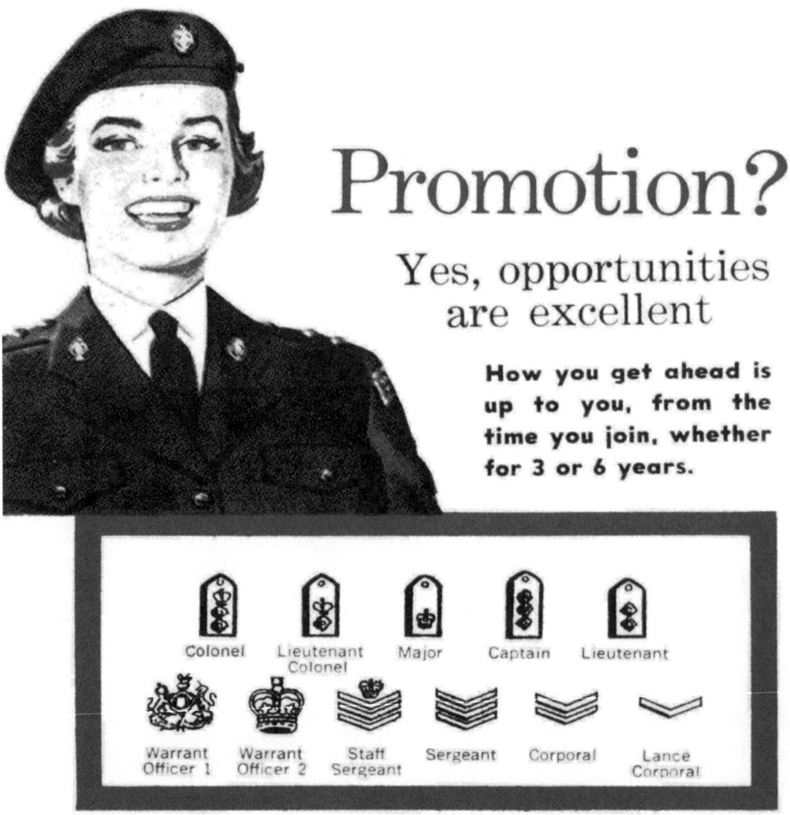

30 WRAAC Barracks at Mt Martha, Victoria is a quiet little area beachside near the Apprentices' School at Balcombe. I was the only female driver in the area. I was not afraid of hard work and when the vehicle, an FB Holden ute, had to be serviced I had to steam clean the underbody and the engine base. I would be kitted up in my dark blue overalls with my beret on my head underneath the vehicle. I was always feminine and wore makeup. We had closed camp on a Monday evening when we were taught to put on makeup and how to be a lady.
Betty Hall (enlisted 1963)

I chose to be a driver as I loved working outdoors. Of course before the driver training at Broadmeadows (Vic.) there was the wonderful rewarding (so not) job of General Duties at the barracks. My first posting as a driver was to 3 BOD (Base Ordnance Depot) from June 1966 to June 1967. I used to drive the other girls from 31 WRAAC Barracks to and from 3 BOD, parking the Combi van in the courtyard at 31 WRAAC Barracks overnight. Occasionally on the way home we would detour via Toorak to look at the beautiful mansions; in the mornings if we felt like a kick start I would drive down behind the sheds (i.e. out of sight) at 3 BOD and do a few wheelies before we started our day jobs. From July 1967 to Jan 1968 I was posted to 101 Transport Unit at Randwick (NSW) and lived at WRAAC School.
Lyn Morris (enlisted 1966)

When I was at Puckapunyal I did two nursing orderly courses held at the Healesville School of Army Health. Both of these courses were hands-on where we learnt to make hospital beds – all the different types of bed-making for different situations, sterilisation procedures, dressing wounds, giving injections, taking temperatures, urine testing and patient welfare. In fact the courses that we'd done qualified us to be fully trained nursing aides in major hospitals. I was at Puckapunyal for forty-three weeks and then I was posted to Southern Command Personnel Depot at Watsonia to the dental unit there where I was a Dental Operating Room Assistant - a DORA. I was the only military person there and I worked with two civilian dentists.
Ruth Grace (enlisted 1964)

Figure 20 Ruth Grace at 3 Camp Hospital, Puckapunyal, Victoria, 1965

Marion Quick was also posted to Puckapunyal.

There were three wards at the hospital plus a theatre which was never used. The wards were quite long, a row of beds down either side, probably twenty-four beds in each ward. The wards were very open, made of corrugated iron, louvre style windows and big verandahs all the way around. The officers of course had their own ward. They wouldn't be seen mixing with the other ranks. The officers' ward was totally different. It had nice furnishings, they had their own ablution block and it was very comfortable down there.
Marion Quick (enlisted 1965)

Margaret Elder became a cook in the Citizen Military Forces (CMF)

Some of the girls that were in the unit that I was in were signallers, drivers and clerks, but I was put in with a group of girls to learn cooking. I don't think we did any cooking on the nights that we went on parade. The weekends were when we did our practical work. As time went on I found out how to do lots of different things in the kitchen. Sgt Riley was a really tough fellow and there was very little hygiene practised. People used to wash their hands, but he used to smoke. There was always a bottle of wine in the storeroom somewhere and the cigarettes would be lying around on the benches while they were doing their work.
Margaret Elder (enlisted 1953)

I was posted back to Sydney, much to my disappointment, and that was to Moorebank Ordnance Depot as a clerk. I settled into barracks at Meadowbank and every day travelled by train to Liverpool West, then by Army truck to work. Truck was the general means of travel and we learnt very early how to hitch the skirt with decorum. After being at the Base Ordnance Depot (BOD) for a few months I was posted to Victoria Barracks as a trainee draughtswoman working with Royal Australian Engineers (RAE). Whilst there the WRAAC Barracks at Meadowbank closed and we were all sent to Georges Heights, Sydney well before it became WRAAC School. It was very interesting as the Artillery unit from down at Middle Head was our security piquet (guarding the area) with many occasions of peering diggers' eyes especially as our ablution block was in the middle of our accommodation block.
Wilma Stralow (enlisted 1955)

When my Clerical Course came up, I had to pack all my possessions and go down to 30 WRAAC Barracks at Mt Martha and live in some horrible little huts with the ablution block away from the huts in the bush a bit. We went to the Army Apprentices School at Balcombe each day to learn to be clerks. After completing the Course, it was pack-up time again and this time I was sent to Puckapunyal to 21 Construction Squadron as the orderly room clerk. On arrival at Seymour Railway Station, I had been told to ring the Squadron and 'they will send someone out to get you'. I had no idea where I was or where I was going. I finally found the phone and it had instructions beside it saying 'turn the handle twice and lift the receiver' if I recall correctly. Of course, I did it back to front and got no answer. Tried it the other way (I was expecting an immediate response) and still no answer. Gave it another go and the operator came on yelling at me to stop turning the handle over and over. I had never seen a phone like it so had no idea what I was doing. After waiting ages outside the station, a Land Rover eventually pulled up with a male soldier who said 'hop in' and off we went to my new posting. I was the only female at 21 Construction Squadron and although I never met him at the time, my husband was also at the same unit. The Squadron was put on standby to go to Borneo and that meant a posting for me. I argued that I wanted to go (not possible for a female), that I wanted to be in the rear party (again not possible) and eventually had to give in, pack up and head for Queensland on the train to 1 Supply Depot Royal Australian Army Service Corps as a rations clerk. I thought I'd hate it after loving Puckapunyal so much but the posting I had was working out the rations for all of South East Queensland units and it was a great job. I really enjoyed the work and whenever the Warrant Officer (WO) Auditor went on leave I got his pay level which was great. Everything was worked out to a ration scale right down to the amount of pepper per person per day. I remember when the dogs were being trained for Vietnam that I thought it was a shame that they were fed the same canned meat week after week and decided they should have some variety. I soon received a phone call telling me to stop sending canned corned beef as the dogs couldn't stop drinking and that wasn't going to help in Vietnam! Hadn't thought of that!
Kay Owen (enlisted 1964)

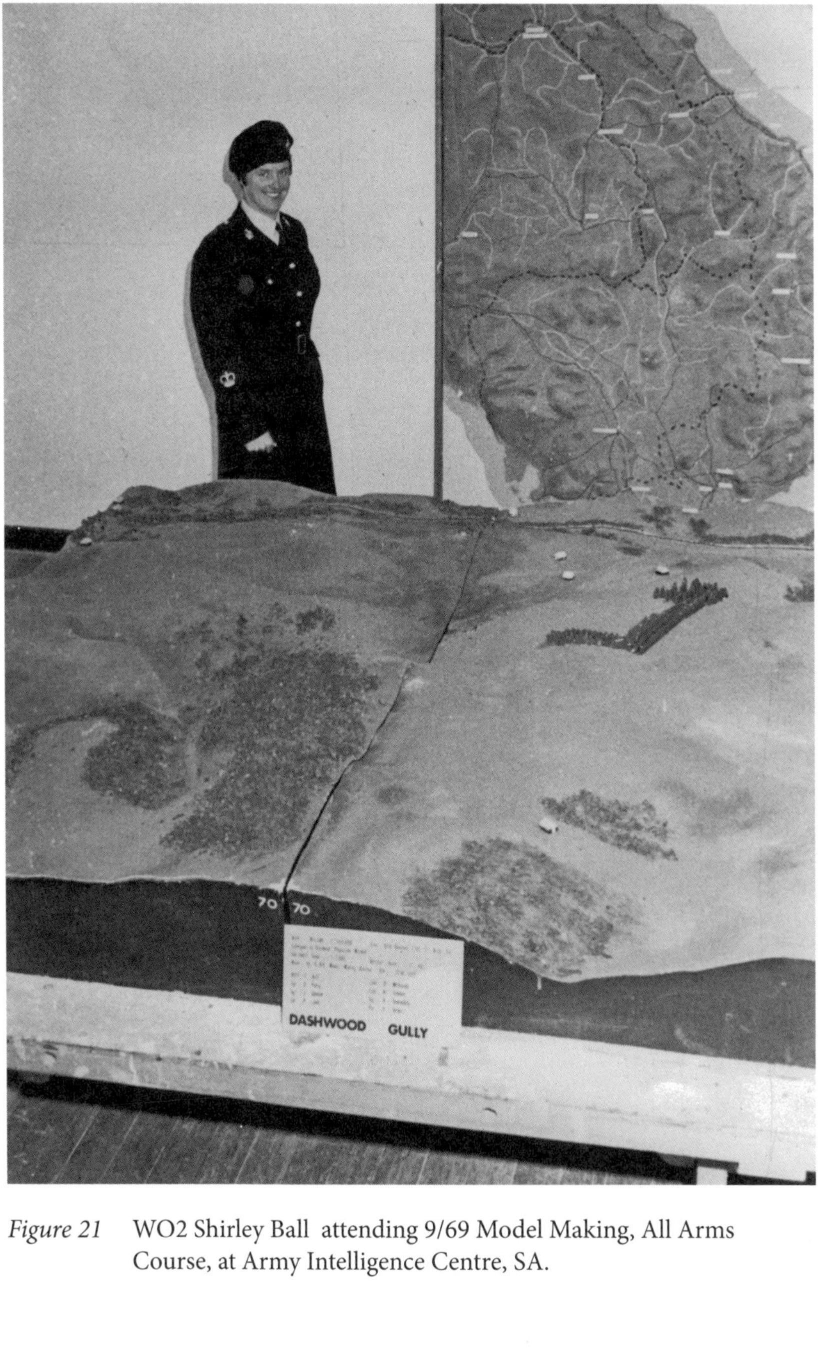

Figure 21 WO2 Shirley Ball attending 9/69 Model Making, All Arms Course, at Army Intelligence Centre, SA.

My first military posting was with the Directorate of Psychology, Army Headquarters (Russell Offices, Canberra). Canberra in the early 1970s resembled a Lego City to me. I had never seen such a clean, well-spaced and planned city before. There were no local shops, corner delicatessens, etc. like I was used to. Industrial areas were separated from residential and retail. I was the only female Other Rank at this posting, working in a small office with a sergeant and several psychologists. The office was located in the same building, on the same floor and next door to the Director of WRAAC. I felt reasonably intimidated by such a large number of high ranking 'neighbours' at work.

During the first few weeks we were working hard on a particular assessment activity. My job was to type the Psychological Examiner results and Psychologist reports. During this period, we were asked what we would like for dinner and I replied that I would like Kenfucky Tried. I blushed with embarrassment at my verbal mistake. I never used bad language. Lt Colonel Goodwin, Deputy Director of Psychology, kindly smiled at my faux pas, but the sergeant was not as kind, teasing me repeatedly about it. During the same week, Colonel Goodwin came in requesting that I correct a report I typed, saying, 'I am sure Sgt ….. does like this hobby, but it should read martial arts, not marital arts'. I was so worried that he and the others would read too much of Freud into these mistakes.

During my first posting, the Director of Psychology counselled me regarding my military career. As I was SG1 (Selection Group Rating 1), he advised that I could apply for Officer Cadet training and that if I would like to pursue this option he would support me. I could not imagine myself in this type of leadership role at that time. I was a child of blue collar migrant workers and I felt intense inferiority with many people. It was not a viable option.
Maria Adamczak (enlisted 1973)

I was told I was going to Toowoomba to be an OpSig (Operator Signals). Loved it. I thought it was very similar to shorthand which I excelled in and I found that Morse code for me was very easy. Haven't forgotten it to this day; in fact I don't know anyone who has forgotten how to read Morse code. (I was posted to) Toowoomba, Darwin, Toowoomba again. We were shiftworkers so we had lots of fun cooking

up a storm on the midnight shifts, particularly in Darwin with the Navy folk.
Sandy Jarrett (enlisted 1980)

After WRAAC School I said, 'Send me anywhere but Victoria.' I was posted to WA and I will have to confess I had forgotten about WA! I had told my mum in Victoria I would see her in six weeks' time!
Jayne Rickard (enlisted 1969)

I was to go to Singapore but they didn't have a posting for a sergeant and were to hold me against a Staff Sergeant posting. My husband was just back from Vietnam. In that time they would send a wife as supernumerary[10] until a posting became available but this time they had to set a precedent the other way around. We were both good at our jobs but our colonel told me that someone was blocking our posting in D (Director) Sigs. This relates to an incident in Vietnam which caused my husband to give info to the Brigadier. Evidence about a chopper taking hits coming into a fire support base just to bring a transistor for an Officer Commanding's (OC's) TV which my husband refused to repair.
Joie Norton-Willson Scheinpflug (enlisted 1967)

In June 1971 I was to become part of ANZUK Sig Regiment which was a combined Australian, New Zealand and United Kingdom group of nearly six hundred men and sixty women from the Army, Navy and Air Force based in Singapore. We were there to contribute to Australia's commitment to the Vietnam War by providing support to our troops by ensuring reliable, confidential communications back to Australia. It was an unbelievable experience working with nine different groups and three different nationalities, both male and female. A true test on some busy shifts. I flew to Singapore on a QANTAS 747 jumbo which was en route to Vietnam. I was the only woman on board as even the hostesses were male. I had an incredible trip up and was bought a drink by half the plane as I would be the last Aussie woman these boys would lay eyes on for at least a year. Arriving at Singapore at 5.30 in the morning the heat hit me in the face like a wet sponge.

I had done six months as a shift leader when I developed a bad allergy to the oil they used on the signal equipment and the ancient teleprinters to keep them operating. This allergy left me with hives and itchy red lumps on any part of my skin that came in contact with the oil on the machines. I had to report to the communication centre leader, Lt Toohey, and explain my predicament. He roared with laughter and reckoned he'd heard every excuse in the book to get out of work but this one took the cake – to actually be allergic to it. In December of 1971 I became his secretary. I often had to accompany him when he met the ships from UK, New Zealand, Australia and USA that came into the naval dockyards at Sambawang. On some occasions we had to ride shotgun for the mail that was delivered to all our bases. This was a wonderful chance to see the island. No WRAAC had ever been trained in the use of weapons so if we were ever to be held up I'm not sure what we were supposed to do. Sitting in the front seat holding a rifle we looked the part although it was so hot in the Land Rover the rifle nearly slipped out of our sweaty hands on the corners.

Libby Flemming (enlisted 1969)

Singapore . . . all expenses paid

Army girls' work is dream

THE dream of most unmarried girls is an oversea "working holiday" before settling down to home and hubby.

While they are scrimping and saving to fulfil that dream, about 90 young women are living it up in exotic Singapore with all expenses paid.

The lucky ones are members of the women's services working with the newly created Australian - New Zealand-United Kingdom (ANZUK) Force.

Australia, as a major contributor to the regional defence pact, has the largest female representation of the three countries.

Of the 91 girls, 48 are with the Women's Royal Australian Army Corps (WRAAC), 22 with the Royal Australian Army Nursing Corps (RAANC), and 21 with the Women's Royal Australian Navy (WRAN).

For them it is a once-in-a-lifetime, all-expenses-paid experience hard to match anywhere in civvy street.

While in Singapore they receive attractive allowances which put their wages way above not only their counterparts at home but also the average Australian working girl.

They are accommodated in modern flat-like barracks, eat three meals a day and commute to and from work, all at Government expense.

So, all that is left for them is to bathe themselves in luxury — Singapore's main tourist commodity.

Duties

With tours of duty on the island ranging from one to two years, these girls have more time on their hands than the everyday tourist and use it to explore the maze of more obscure, less expensive shops.

Apart from the odd clerk or cook, nearly all the WRAAC and WRAN girls work with the ANZUK Signal Regiment which is unique because it is tri-national, tri-service and dual-sexed.

Perhaps the biggest and most beneficial change in working conditions for the girls in Singapore has affected the nurses and nursing sisters of the RAANC.

At the 150-bed ANZUK Military Hospital, comparable to a full civilian general hospital, and the 38-bed ANZUK Woodlands Hospital, a minor care hospital and families clinic, these nurses are caring for wives and children as well as servicemen, possibly for the first time since joining the Army.

WRAAC Cpl Elizabeth Flemming, of Devonport, Tas, turns up a wide-brimmed coolie hat at a bargain stall in Singapore. It is her first time out of Australia and she is making the most of every shopping opportunity she gets.

Figure 22 Newspaper article featuring Libby Flemming in *The Mercury*, Hobart, Wednesday 26 April 1972

I was very lucky in being the first woman to do the Nuclear Operations Instructors Course and the Biological and Chemical Defence Instructors Course. That was exciting. I met with quite a lot of resistance from my male counterparts, not necessarily my fellow students but more with the instructors on the course. I became affiliated with the Corps of Signals and was very grateful for the opportunity to go on an overseas exchange posting to England to 30 Signal Regiment in Blandford Forum, Dorset for two years. I was competing with my male counterparts so I was pretty happy to get the opportunity to go over there. So here I was, the first female officer in the Australian Army to go to the exchange posting for Signals Corps. I thought I was going to be the Assistant Operations Officer but as it turned out I was going to be the second in command of one of the squadrons. The regiment had recently returned from the Falklands so what I had was about 220 young soldiers that were ten foot tall and bulletproof who had just come back from operational service.
Denise Dunn (enlisted 1976)

My first posting was to Singleton Transport Unit in the Hunter Valley from about August 1975 to December 1976. I was one of two women in the unit, the other being a driver. My Chief Clerk gave me a variety of work and also encouraged me to qualify as a unit pay representative. A pay rep manually entered the pay entitlements of each soldier into a separate pay book and then transferred the detail onto a large sheet. The large pay sheet, fitted with carbon paper, would be totaled and the amount would then be requested from the Cash Office for delivery on the morning of pay day. A breakdown of all the money you needed, to make up each individual person's pay, had to also be given. Once the money was delivered you had to check the entire amount against what you had ordered, e.g. 100 x $20 notes, and then you would start counting it out into individual pay books. At the end you were not supposed to have had anything left over. If you did, you had to check every book again until it was correct.

All pay reps were required to have a 9mm pistol in case they were robbed and I was given an unloaded one because I had never been trained to use it. I was told to simply hand the money over if a robbery did occur. Different junior officers were assigned to conduct pay parades. The routine was to count the entire amount, then count it into individual pay books and achieve a zero balance. One young officer decided to

count the total amount and then pay each soldier as they turned up from the total fund believing it would all work out. Unfortunately he ended up short at the end of the parade. A tough lesson learnt. After my promotion to Staff Sergeant I was offered a position in Darwin at the same rank. At first I found it difficult to maintain the required level of fluids needed to be hydrated but I loved being in a camouflage uniform every day and not having to deal with the southern winter. I was part of the Ready Reaction Force if trouble broke out in Indonesia but it didn't during my posting term. When I first arrived there were only about 60,000 people in Darwin and the military numbers were small. It was a credit to the many Army Reserve members who filled the positions and kept Defence going. By the time I left, Robertson Barracks was being built and I was fielding many phone calls about my position and when it would be likely for me to move on. I admired all the Army Reserve people who not only had a full life away from the military but were also quite dedicated to the military. It was great to see so many women with lengthy service and how well they were respected within the unit.
Christina Morgan (enlisted 1975)

There was a Corporal pay position vacant at 3 Camp Hospital, Puckapunyal and that's how I got into the pay stream from clerical. I got promoted to corporal in that job while I was pregnant with my first child. They introduced maternity leave around that time so I took the three months maternity leave after I had Anthony and I went back into my job. Four years later I was posted to Randwick and Peter was posted to Holsworthy, NSW. We lived in Holsworthy married quarters and I commuted to Randwick on a daily basis. My husband, Peter, looked after Anthony, got him to the babysitter and picked him up after school. Peter never held me back from promotion; he was very, very supportive. He saw that as being a very good thing for the family. I had very good bosses as well and I didn't experience any prejudice. I can honestly say I had a very comfortable career. I was very lucky. I had another child in 1987 when I was in Randwick, and from there we got posted back to Puckapunyal because we set my career as a precedence (i.e. gave priority to my career). Previously, wherever the man got posted to, if there was a posting for the lady he'd go. But times were definitely changing as I was progressing to the point where I was able to say to my career advisor 'My career has precedence'. My husband after twenty-five years retired as a corporal and he was happy with that rank. I reached the rank of Warrant

Officer Class II. He didn't want to progress any further because he didn't like the culture of the sergeants' mess where you would go in and the doors would be locked and you had to stay until the Regimental Sergeant Major (RSM) said you could go. Even when I went to the mess I would keep low-key. I would drink water and as soon as it was time to go I was out of there.

Anna Celi (enlisted 1976)

LIFE IS GOOD IN THE WRAAC

Accommodation: The Army provides all accommodation and meals in addition to the pay and allowances quoted in the Pay Leaflet. In the WRAAC Barracks in the capital cities or in the WRAAC Quarters in units throughout the country you will find that the Army maintains the highest standards of accommodation. Bedrooms are attractively and comfortably furnished and you may add your own personality to your room with pictures, photographs and flowers. Dining rooms are pleasantly decorated and entertainment facilities usually include recreation room with T.V., radiograms and piano plus a lounge for the entertainment of friends and visitors.

Chapter 4
The Barracks

The standard of the female soldiers' living quarters ranged from very basic to quite up-market. The accommodation for recruits at WRAAC School at Georges Heights, Mosman, Sydney, consisted of fibro and iron buildings opening onto verandahs which provided a covered walkway to the ablution blocks which were situated between every two huts. The rooms contained four metal-framed beds and a wardrobe and metal-framed desk and chair for each person. All the furniture and the linoleum on the floors was brown.

The women's quarters at Kapooka, near Wagga Wagga, NSW, were very similar though on a much smaller scale, as were the quarters at Campbell Barracks, Swanbourne, WA. However other women's barracks were much more comfortable. 30 WRAAC Barracks at Mt Martha, Victoria, was a pretty old weatherboard and iron one-storey house near the sea. The women's barracks in Brisbane was a picturesque old former Army Hospital overlooking the Brisbane River though the conditions were not exactly luxurious, with the women having to tend the boilers to get hot water for showers and some having to sleep on verandahs. In Hobart the women were accommodated in the historic Anglesea Barracks which was built in 1814, and in Adelaide their quarters were in Keswick Barracks built in the early years of the 20th century. In Melbourne the women were accommodated at 31 WRAAC Barracks, a stately old Victorian building in St Kilda Road, a ten-minute tram ride from the city. These and other women's barracks are described in the stories in this chapter.
Lyn Kane Dale

Figure 23 31 WRAAC Barracks, Melbourne, Victoria

Figure 24 Keswick Barracks, Adelaide, SA

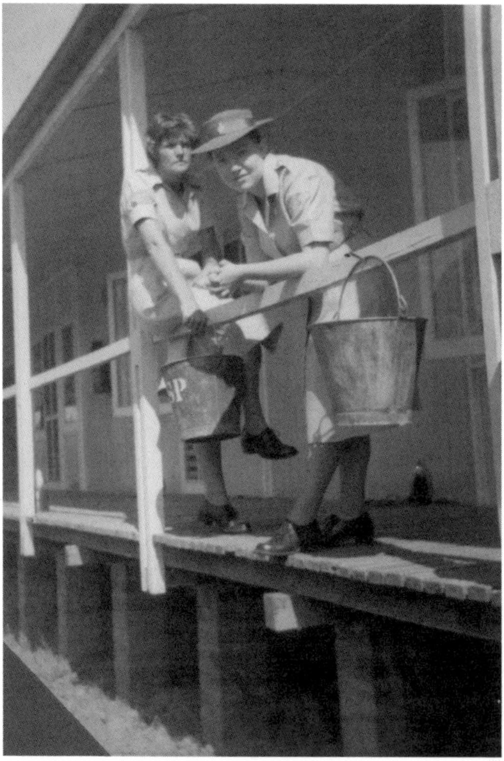

Figure 25 GDs at female quarters, 7 Camp Hospital, Kapooka, NSW, 1964. Vera Alexander (left) and Lyn Kane.

In early 1965 I finally got my clerical posting and left Kapooka at 4 am to go to my new unit, 3 Base Ordnance Depot in Melbourne. The train from Wagga Wagga arrived at Spencer Street station in Melbourne midmorning. At the baggage counter I saw a familiar green WRAAC uniform and the driver came up to me, introduced herself as Fran Farrell and helped me carry my cases to the vehicle. We had a short drive through the busy city and then turned into St Kilda Road, which looked so beautiful and green after dry and dusty Kapooka. The street was lined with beautiful old trees, lovely flower beds along the median strips and wonderful old stately two-storey Victorian mansions with magnificent gardens. We drove for several miles along St Kilda Road and then turned

into a back lane and in through a gateway and I realised that my new home was one of these lovely old buildings. It certainly was a beautiful house with wooden floors (plenty of polishing to be done there!) and high ceilings. There were approximately 100 residents at any given time – someone was always arriving or leaving. The sergeants, who were a fairly small group, lived in the former servants' quarters above the old stables, which had been converted into laundries and drying rooms. In the concrete covered courtyard were several Hills hoist clotheslines and there was room to park several vehicles. The officers' quarters were upstairs in one wing and the other ranks occupied the rest of the first floor and one wing of the ground floor. Also on the ground floor were the three messes for officers, sergeants and other ranks, a common room, Regimental Aid Post (RAP), other ranks' canteen and kitchens and storage rooms. There was a bathroom for every couple of rooms so five or six people shared one bathroom.
Lyn Kane (enlisted 1964)

My home for the next three years would be 31 WRAAC Barracks, 478 St Kilda Road, Melbourne, and what a wonderful three years it was too. 31 was certainly different from our billets at WRAAC School and other Army barracks. Previously a hotel, a grand building, cloisters out the front, jarrah staircases, French doors to the dining rooms, stained glass windows featuring Australian birds, lovely lounge room that was perfect on a cold wintry day to read a good book in front of the heater. The internal courtyard where a sunny day would see the girls come out to sunbake. Our rooms had two, three or four beds each with a bathroom in between. Our social life was nearly always good to great. There was always something going on. A phone call from another mess meant a party. The best party for me was the Harlequin night. We all dressed up in crinolines and I invited some guys in the Provost Corps who were heading off for a stint in Vietnam. What a great night – danced all night. Even our commanding officer, Captain Reynolds, affectionately known as 'Debbie', joined in the festivities.
Cherryl Burrows (enlisted 1964)

It was very nice there. The food was good, the company was good. I used to have the weekend with my aunty and cousins and then she'd take me back on Sunday night which was great, so I wasn't too lonely in 31 WRAAC Barracks. I just liked the camaraderie, the girls, the discipline,

the marching and the uniforms. It was just a great life. I was living on my own when I was younger so having that sort of family around me was very good.
Desiree Harms (enlisted 1965)

Figure 26 Army publicity photograph of Desiree Harms and a sailor outside the National Gallery of Victoria, c. 1968.

My five and a half years at 31 WRAAC Barracks.

I completed my driving course on 7 August 1963 at Infantry Centre, NSW and was posted to 403 Signals Regiment, Watsonia, but the posting was changed overnight to 31 WRAAC Barracks. I arrived at 31 August 1963 and departed 24 December 1968.

My duties as Driver were:

Understudy to everyone and everything in the Orderly Room.

Switchboard – answering the phone and paging who the calls were for.

Re-addressing mail and keeping the re-address book up to date.

Restocking the canteen, checking the stock and money.

Picking up the rations every day from the depot in South Melbourne and once a week filling the vehicle up with fuel.

Tuesday was laundry day and I'd help pack the baskets with the dirty linen.

Every Thursday I would go to Australian Services Canteens Organisation (ASCO) at Maribyrnong to pick up canteen orders for the messes and unload them on my return.

About once a fortnight I'd go to a deli in South Melbourne to pick up the officers' cheeses, etc.

It was not all work and no play. From 1964 to 1967 we had five official parties in the Other Ranks (ORs) mess organised by the Mess President and her assistants. These were:

Christmas Party on Saturday 5 December 1964

Christmas Party on Saturday 4 December 1965

Roaring Twenties Party on Saturday 27 August 1966

Christmas Party on Saturday 3 December 1966

Harlequin Party on Saturday 3 June 1967

We had the best of cooks (but) in 1967 the WRAAC seemed to run out of cook replacements, so HQ Catering, Melbourne, employed two male cooks for the barracks. The trouble was they came from different countries and had disagreements. It came to a head one day when one cook locked himself in the men's toilet and the other was outside with the carving knife. Lieutenant Gorman and I, along with our young Vietnam

veteran (seconded to the barracks), got him in the Adjutant's office to find out what was going on when he turned and head-butted the Vietnam vet, dropping him to his knees, and fled out the door through the back gate never to be seen again. That was the end of the infamous male cooks. The General Duties (GDs) kept the barracks spotless from top to bottom every day, polishing, dusting the officers' quarters, sergeants' quarters, anterooms, hallways and staircases. The stewardesses looked after the officers' and sergeants' messes and kept the ORs mess and pantry spick and span. The girls came and went after each recruit course passed out of WRAAC School and those who were awaiting their respective training courses departed. In the early part of 1967 I decided to change to clerical after talking to Captain Reynolds (Officer Commanding the barracks). She arranged for me to be put on Clerical Course from June to August and then I was posted back to 31 WRAAC Barracks and promoted to Temporary Corporal in the posting of Clerk/Storewoman. I did my Non-Commissioned Officer (NCO) course at WRAAC School from January to March 1968 and then I was promoted to Temporary Sergeant as the Orderly Room Clerk. In August my sister died and my mother had a heart attack. I then put in for a compassionate posting to NSW, close to home, so I could travel to work from home. I left 31 WRAAC Barracks on 24 December 1968 for HQ Liverpool, NSW. My posting was reviewed in 1970 and as I only had twelve months to serve they carried my compassionate posting for a further period and I was posted to 2 Military Hospital as Admin and Discharge Sergeant. I was discharged from the Army on 16 December 1971.
Fran Farrell (enlisted 1963)

After WRAAC School my first experience of barracks life was living at 31 WRAAC Barracks when I was posted to Albert Park Barracks, Melbourne. What a wonderful experience of having a fantastic group of young women and officers around me.
Denise Darbey (enlisted 1973)

I was transferred to 31 WRAAC Barracks where my world opened up and my social life was complete along with all my big sisters. We could go out and have a good time and felt safe knowing we would look after each other. It was like a big family and I felt very secure.
Ann Dullaart (enlisted 1966)

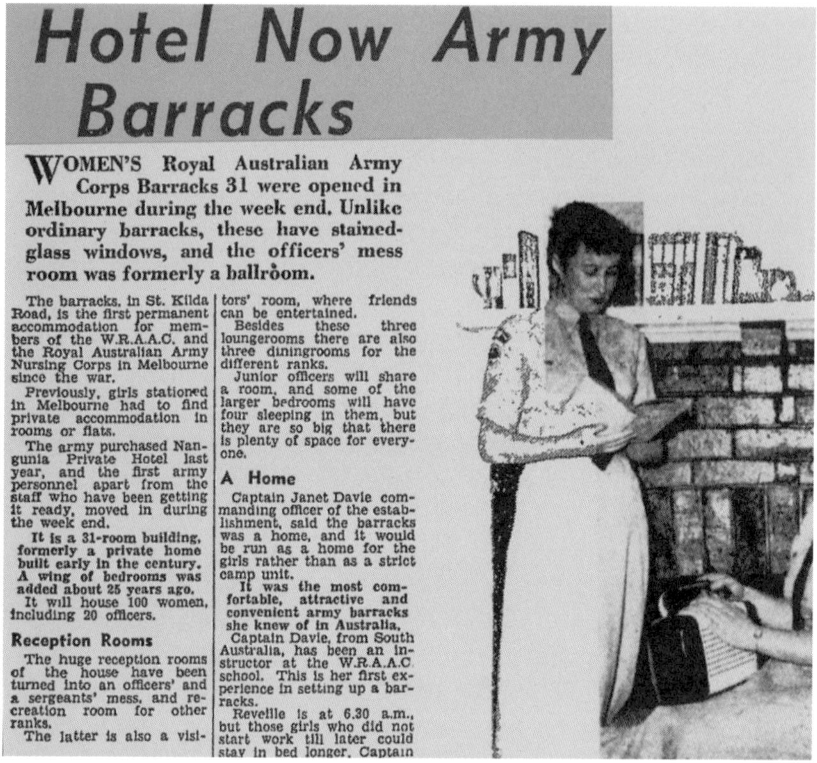

Hotel Now Army Barracks

WOMEN'S Royal Australian Army Corps Barracks 31 were opened in Melbourne during the week end. Unlike ordinary barracks, these have stained-glass windows, and the officers' mess room was formerly a ballroom.

The barracks, in St. Kilda Road, is the first permanent accommodation for members of the W.R.A.A.C. and the Royal Australian Army Nursing Corps in Melbourne since the war.

Previously, girls stationed in Melbourne had to find private accommodation in rooms or flats.

The army purchased Nangunia Private Hotel last year, and the first army personnel apart from the staff who have been getting it ready, moved in during the week end.

It is a 31-room building, formerly a private home built early in the century. A wing of bedrooms was added about 25 years ago.

It will house 100 women, including 20 officers.

Reception Rooms

The huge reception rooms of the house have been turned into an officers' and a sergeants' mess, and recreation room for other ranks.

The latter is also a visitors' room, where friends can be entertained.

Besides these three loungerooms there are also three diningrooms for the different ranks.

Junior officers will share a room, and some of the larger bedrooms will have four sleeping in them, but they are so big that there is plenty of space for everyone.

A Home

Captain Janet Davie commanding officer of the establishment, said the barracks was a home, and it would be run as a home for the girls rather than as a strict camp unit.

It was the most comfortable, attractive and convenient army barracks she knew of in Australia.

Captain Davie, from South Australia, has been an instructor at the W.R.A.A.C. school. This is her first experience in setting up a barracks.

Reveille is at 6.30 a.m., but those girls who did not start work till later could stay in bed longer. Captain

Figure 27 WRAAC Barracks, Mt Martha, Victoria

The Haunted Cottage. I was attending a clerical course at 30 WRAAC Barracks, Mt Martha, Victoria, and six of us girls were given the 'Cottage' as our accommodation. Strange things happened to us all there. When we arrived, Sue Davis took out an empty wine bottle and put a candle in it. The next day after attending our class all six of us were called up to the Commanding Officer (CO's) office, where she accused us all of drinking in the Cottage. She said the smell of wine was very strong when they did their inspection. Sue owned up that the bottle was hers, but we all denied drinking, telling her the bottle was empty upon arrival. It didn't matter. We all got Confined to Barracks (CB) for a week, just in case! I was alone in the cottage one night and will still swear to this day, that someone or something was heavy breathing over my shoulder. I stuck my head under the blankets and hoped for the best! Another girl awoke one night to see what looked like the bottom half of a man, from the waist down, hanging from the back of her door. There was nothing hanging on the door in the morning. My room mate, Patty Newel, was at the Cottage at a later date for another course and mentioned that she saw something white fly past her window. There were other incidents, but I can't recall them all.
Jean Fullerton (enlisted 1963)

The original WRAAC Barracks at Moorebank, NSW, were old huts with a tin fence around us, and a gate that was locked at the end of leave hours. You had to apply in writing for a leave pass to go out at night, and Monday was closed camp, for lectures, etc. The camp was in a parkland setting with a golf course across the river and the Army depots up the other way. We were right in the middle of the camp area, with the duty room (men) across the road and no inline phone. Any calls for the girls came through the duty room with the men sitting there listening. We eventually had a better phone system but it was manned by the diggers at School of Military Engineering (SME) and they always listened in. One night a naked drunken digger was thrown over the fence. Poor devil was posted out very quickly. He was so drunk he had no recollection of the whole night!! After maybe two or three years a new single storey brick building was built for us with four to a room including corporals. There was no fence this time.
Merle Viles (enlisted 1962)

In mid-1970 I was stationed at Bendigo and was tasked with doing GDs. I spent a couple of months there before going on course. The barracks for the females was in a lovely old two-storey house, with a donga[11] for the sergeants out the back and a tennis court where we held morning parades. The house had been converted for us with (our) accommodation upstairs. The Major had her dining room and quarters downstairs along with the kitchen, dining room and lounge/rec room. There were approximately twelve of us living there at this time. The cook, Karen Byrne, and I were left on our own during the day with only the duty driver arriving each day with our rations from Fortuna. Once a week we went to the main mess for lunch. The boozer (canteen), where the girls were allowed to have male guests, was in the cellar with leadlight windows at ground level – very dark and mysterious.

I didn't spend much time at Fortuna but the cartography section worked in the old ballroom. Topo (topography) had a camera with a lens large enough for someone to stand inside – if they were allowed!
Hilary Jones (enlisted 1970)

Figure 28 Home of the Army Survey Regiment 1942-1996, Fortuna, Bendigo, Victoria

All of the women, despite where they were posted within Singleton (NSW) were accommodated in the same barracks. These consisted of two wooden buildings with a grassed area between. One building held the privates and corporals and the other held a RAANC corporal and one officer. The rooms were for two people although some were for four and you could move your furniture around. The furniture consisted of a single bed, desk, chest of drawers and a wardrobe. Many used their wardrobes to create divisions in the room for privacy. In the middle of our building was a recreation room consisting of a complete kitchen and lounge area with a black and white television. The showers and toilets were next to this and then there was a laundry area for us all to use. To access the common areas you had to leave your room and walk along an open verandah. In winter you would often get wet going to the showers. On a Tuesday night it was 'Closed Camp' which meant no one could go out. You were supposed to be cleaning your room and ironing your uniforms. The most senior WRAAC was known as Madam WRAAC and she had the power to discipline any WRAAC on site. On a Tuesday night she would sometimes get the women to kneel in front of her to measure the length of the skirt from the knee as miniskirts were very popular and some of the women had taken the hem of their dresses up. Each Wednesday she would inspect our rooms before we could head off to work. The mess was in another building and we ate with the men. Some women opted to eat in the recreation room instead after having prepared their own food.
Christina Morgan (enlisted 1975)

In those days in Canberra there were no barracks for female soldiers, so we were placed at the Gowrie Private Hotel, Northbourne Avenue, Braddon (ACT). It was a large private hotel that accommodated public servants. It was affectionately nicknamed the 'twin towers of sin', as there were two wings, both having approximately nine floors, with cafeteria, television rooms, etc. on the ground floor that joined the two buildings. It was wonderful as all rooms were single ones with hot/cold water basin and serviced, so no cleaning of rooms and making beds to military standards.
Maria Adamczak (enlisted 1973)

If you didn't go out and rent a place of your own in Canberra, you were boarded at Gowrie Hostel, as there were no barracks as such. This wasn't so bad, and nearly everyone respected the fact that there were shiftworkers who kept really odd hours. You could always find someone who was having a party somewhere on the floors.
Clare Williams (enlisted 1980)

I loved the camaraderie of barracks and mess living. After WRAAC School I was posted to Perth and twelve of us lived in the nurses' quarters at the hospital of the Special Air Service Regiment. There were 400 blokes there and another 200 overseas in Vietnam. At one point five out of twelve (girls) in the quarters were pregnant. One of the girls had a concealed pregnancy and went into labour at work at Swan Barracks. We were all ordered to 'go on the pill'.
Jayne Rickard (enlisted 1969)

Anglesea Barracks, situated in Davey Street, Hobart, was a great place to be posted. We enjoyed a wonderful social life at our OR's Club, the Magee Club. We were one big happy family. Anglesea Barracks is recognised as one of Australia's most significant historical military precincts and its appearance today is much as it was when the first buildings were constructed in 1814. What was once the Military Gaol is where my Pay Office was situated, the gaol having been built in 1847. WRAAC very rarely went to the ORs mess for dining. Within our barracks we had a small kitchen which we used. One of the girls used to have her breakfast when she arrived home late at night. She claimed it saved her getting up early for a meal in the morning. I well remember the night one of our National Service Engineers thought he would liven up our lives by doing a streak through the barracks which caused a riot and a lot of accusations being made. It was questioned whether he was doing a streak or in fact had been with one of the WRAAC. A very distraught time was in November 1970 when our main WRAAC Barracks was engulfed in fire and resulting in the death of one of our WRAAC. I was to have been Duty NCO that night but I had a roster change. My husband was in Vietnam at the time. He had a message of the fire but did not know for days that I was okay. At this particular time I was living out, having been married in March that year.
Andrea Woods (enlisted 1967)

Figure 29 Andrea Woods at Anglesea Barracks, Hobart, c 1969.

Life in the WRAAC Barracks in Brisbane was better than in Melbourne. Living dormitory style was not great but the girls were good fun. As usual, we had our turns at polishing the wooden floors with wild and almost uncontrollable polishers. We had mosquito nets over our beds and on waking up in the mornings, would find enormous rhino beetles all clinging madly to the nets. The first night I was there, possums in the roof started fighting and I woke up terrified thinking we were being attacked or something. Everyone else just said 'Go back to sleep, it's only the possums!' When I was rostered as Duty Corporal, I would have to stoke the old boiler that was down the back of the barracks. We had to walk through long, unkempt grass into the boiler shed and shovel coal into the boiler to keep the hot water maintained. This was done in full uniform and not a clean job to do. The spiders and imagined wildlife in the grass during the night made it a frightful job.
Kay Owen (enlisted 1964)

Figure 30 WRAAC Barracks in Brisbane, Queensland

The female quarters in Brisbane used to be a Red Cross Hospital through World War II and resided on the banks of the Brisbane River at Chelmer – very old, we had dormitories. There were two verandahs and I was out on the smaller of the two. We had to keep the fire burning so that we could get hot water for showers and cooking and things like that, particularly on a weekend. If the fire went out and someone had a cold shower we well and truly knew about it and of course the cooks relied on it for the hot water for their kitchen duties. We'd have possums running across the roof of a night time. We found a snake out the back one day.

We had an outdoor loo – every day they came around and emptied it. We knew what time the truck was coming and any new staff member who didn't, when they first arrived we didn't tell them not to go at 10 o'clock in the morning because they'd be confronted with the truck arriving. Not very pleasant, but still. We were right next door to an orphanage. At Christmas time we would put on a Christmas party for the children. We'd have the old three-legged races and spoon races and all those sort of things that were, you know, in those times. We'd have a Father Christmas come and we thoroughly enjoyed it. I'm sure the children enjoyed it as well. I was on duty over one Christmas and we had the traditional Christmas meal. We may have complained about the cooks throughout the year but this Christmas they really excelled themselves. We had the officers waiting on us and it was all very nice.
Christine Negus (enlisted 1963)

I lived in the nurses' quarters at 3 Camp Hospital, Puckapunyal, Victoria, and I must say I loved living there. The 'powers that be' eventually decided the WRAAC should be in separate quarters and built some horrible 'modern' huts beside the old quarters and we lost the 'homey' rooms with wooden furniture and the chance to 'decorate' our rooms with personal possessions. I even had a cat and a budgerigar which had to go to someone staying in the old quarters. The new rooms had metal furniture and one person per room.
Kay Owen (enlisted 1964)

If you wanted to move out of the accommodation provided, into the community, you had to apply and this would be considered. Other women living-in, predominantly from 21 Supply Battalion, told of the senior female officer of their unit inspecting the property to ensure it was

suitable. Your application to live out could be withdrawn at any time as the premises could be inspected by a senior WRAAC officer and you would be ordered to move back in. I was told there were women back in the live-in lines due to a lack of food in the fridge or men's clothing being found in the rented premises. Despite such warnings I moved out with a female Military Police (MP) Corporal. My unit contacted hers to make sure we were moving out together. We both had bean bags, I had a stereo record player and she owned a television. This we thought would be enough, although we did quickly acquire beds and a fridge. Little did I know that I would soon be on the move again. One of the female officers at School of Military Engineering (SME) lobbied the CO to allow unit females to be accommodated at the unit but to no avail. She told us that she said, 'But we are all adults' in support of being accommodated together and the CO responded with, 'Exactly. You do see my point' and continued to refuse to have females accommodated at SME.
Christina Morgan (enlisted 1975)

Figure 31 Gloria Coombe, Puckapunyal, Victoria, c.1961

joie de vivre

noun: **joie de vivre**

1. exuberant enjoyment of life.

   'they seem to be filled with joie de vivre'

   synonym gaiety, cheerfulness, cheeriness, merriment, light heartedness, happiness, joy, joyfulness, joyousness, delight, pleasure, high spirits, spiritedness, jollity, jolliness, joviality, exuberance, ebullience, liveliness, vivacity, enthusiasm, enjoyment, verve, gusto, relish, animation, effervescence, sparkle, buoyancy, sprightliness, jauntiness, zest, zestfulness

   'Free online dictionary'

Chapter 5

Happy Days

One of the advantages of living in barracks was that something interesting was always happening. At 31 WRAAC Barracks in Melbourne, girls who might not have had a date on a Saturday night could find themselves going in a group to a party or out for a meal. Many young women of the 1950s and 1960s had never dined out before they joined the Army, and Melbourne had so many wonderful restaurants to enjoy. The women were invited to parties, dances and balls and went out in groups to the theatre and to movies. Most had a very full social life. My room mate, Shirley Hall, and I often went to dances at the Victorian Railway Institute (part of the Flinders Street station building) on Saturday nights.

Melbourne was the clothes capital of Australia and many young women enjoyed shopping for casual clothes or beautiful dresses for evening wear. There was always someone to go shopping with after work or on a Saturday morning before the shops shut at midday. For the two years I spent in Melbourne a large proportion of my holiday pay was spent on lovely clothes to take home to Perth to wear at Christmas – dresses, pants and tops with matching shoes and bags, to say nothing of beautiful underwear and sleepwear.

Confidences were shared while sitting in each other's rooms, spit polishing shoes or ironing uniforms while listening to music on portable record players or transistor radios. Many of the girls had Birkos (an open top electric kettle), and they could cook soup, boil eggs or heat up tinned food in their rooms, which of course was strictly forbidden. Each girl was issued with an electric bar heater and these were often used to cook toast on. On weekends especially, anyone walking along the corridors could smell toast cooking. There was a communal TV room in all barracks so girls would sit in the TV room together munching chocolate or chips and enjoying whatever was being screened. On the few hot, sunny days of Melbourne summers, girls would lie on beach towels in the courtyard of the barracks hoping to get a suntan.

Letter writing played quite a large part in the lives of these young women in the 1950s and 1960s. In the days before emails and mobile phones, interstate telephone calls were expensive and entailed getting enough coins to keep feeding into the slot in the telephone box when phoning home. Making life a little more awkward was the fact that not all homes had telephones so arrangements sometimes had to be made for parents to go to neighbours' or relatives' houses to phone their daughters. In my case, my parents, brother and three younger sisters would go to the next door neighbours' home to phone me on special occasions like my birthday. Letters were a much more efficient and private experience as the letter writer could just concentrate on what she was going to write rather than stand in a phone box, often with others waiting outside for their turn to use the phone and eavesdropping on the conversation. Many of the women wrote home to their families on weekends. My mother wrote to me twice a week for the five years that I was in the Army. It was always a wonderful feeling to come back to the barracks and look on the noticeboard to see a letter addressed to me. I kept all the letters that Mum wrote to me but on my final trip home found that I didn't have enough room in the suitcase for the large bundle of letters and so I burned them, something which I have always regretted.

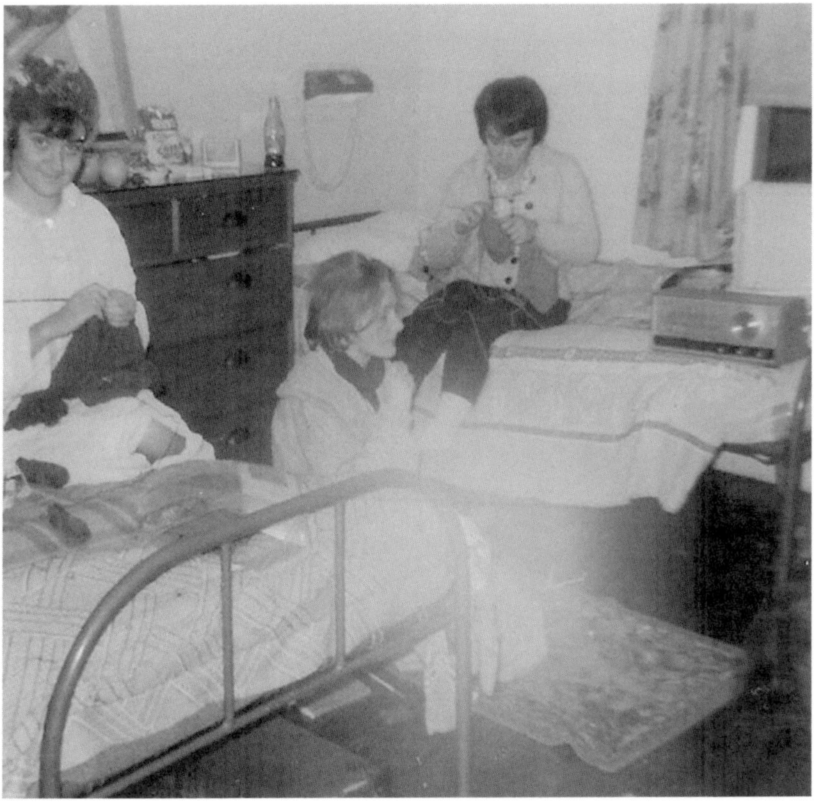

Figure 32 Jean Fullerton (far left), Shirley Hall (far right).
31 WRAAC Barracks, Melbourne, c.1964

Many of these young soldiers had never been far from home before and they took advantage of their free time to explore their surroundings and to go further afield. On weekends small groups from 31 WRAAC Barracks went on bus trips to the old goldfields towns of Ballarat and Bendigo or to the snowfields at Mt Buller, Victoria, where the Army had a chalet where Army personnel could stay for $1 each per night. It was possible to book rooms but any latecomers had to bunk down together and sleep on mattresses in the loft. Some of the women visited the Snowy Mountains (NSW), Canberra, Mt Gambier (SA) and other beauty spots. I was in one group of four WRAAC who caught the 'Princess of Tasmania' across Bass Strait to Devonport on a very stormy night in September 1966 for a four day long weekend in Tasmania. We were unable to get a cabin on the way over and had to sit up all night. The sea was very rough and, to avoid seasickness, a number of people stayed on deck all night even though it was bitterly cold. We then caught a bus to Hobart where we stayed in a very small, very old hotel. We hired a car and driver for $20 per day - $5 each, and he drove us to Port Arthur and other historic places and beauty spots, and kept up a running commentary on places of interest. When it was time to go back to Devonport to catch the ship back to Melbourne, instead of again taking the bus, our driver drove us all the way back to Devonport and took us right to the docks.

All of these experiences, as well as being given jobs which required initiative and commitment, were instrumental in enabling these young women to mature and become independent - skills which would be of great value to them in their future lives.
Lyn Kane Dale

On completing my basic training at WRAAC School, my first posting was to 3 Camp Hospital at Puckapunyal (Victoria). I arrived there on a Saturday morning with another girl. As we walked into the mess we were greeted by a number of people, most of them WRAAC rather than Nursing Corps. We were shown our rooms and were invited back down to the mess for lunch. While we were having lunch one of the WRAAC approached me and asked if I would be interested in going up to the snow fields that afternoon with her and a friend. I had never been to the snow and I thought that would be nice but I was a bit concerned about how far away the snow fields were and whether we'd be able to be back to base before curfew. She said we would and not to worry about it so I ended up going. It was my first experience with snow. I had a nice time and there were no problems. We were back on time.
Marion Quick (enlisted 1965)

Figure 33 Shirley Hall on tour, 1966

While posted at WRAAC School I went to movies, shopping, trips on the ferries to Bondi and enjoyed going to Taronga Zoo. During the summer we often went down to the local beaches on our time-off. Twice at WRAAC School, I was involved in TV activities. The first was as an extra for *The Bobby Limb Show*, a variety programme where they featured a wartime theme. I was in uniform and danced while the songs were sung. I never saw the show so am not sure how it looked. The second activity was for a promotional show for the Army, especially the WRAAC. We were on the basketball court at WRAAC School and the cameras showed us having a game and winning! On the work side, I was taking coffee into the officers' lounge room one night when the tray tipped and coffee went all over the carpet! The person who helped calm me and

suggested what to do was, of all people, the Officer Commanding (OC) at WRAAC School.
Daphne Gardner (enlisted 1964)

Road safety was not what it is today and it was nothing to travel from Singleton to Brisbane, Sydney, Canberra or Melbourne by car for a weekend. On one such trip to Canberra, the car went off the road and plunged some distance down. I was the only one who sustained an injury - a cut to my chin. This was rather fortunate as I had not been wearing a seatbelt. I had been wedged into the back with everyone's clothes (that) they were taking for the trip away which cushioned the impact for me. I was lucky.
Christina Morgan (enlisted 1975)

It wasn't until I joined the Army and was posted to Victoria that I ever set foot in a restaurant. My, oh my, I felt really posh! Restaurants were few and very far between in WA in those days. In fact when I left home in 1963, there were no memories of any restaurants at all tucked into my memory banks. Perhaps it was just that my family could never have afforded to eat at a restaurant. A meal out in those days was to a milk bar for steak, chops, sausages and chips, a hamburger, or fish and chips wrapped in newspaper, and eaten down at the beach. Melbourne introduced me to a new level of sophistication, and, when one of my roommates, June Aulfrey, and I, went to the Southern Cross Hotel restaurant in 1964 for a special 'farewell' tea (the West Australian word for dinner), we took great delight in shopping for new frocks, complete with matching shoes and handbags. We had moved on at this time and no longer thought it necessary to have matching gloves. We were sophisticated ladies. The word 'ladies' lent an air of gentility and refinement. Of course, now we wouldn't say ladies, we'd use the term 'women'. In 1964, the term women would have been slightly derogatory. It's amazing how times have changed.

Saying farewell to June, whose six-year enlistment was finishing, was my first experience at losing friends to postings or discharges. June, from Emerald, Qld, had been a jillaroo before enlistment to do nursing in the Army. A double-date with June was always an eye-opener for whatever lads we were dating when June would rattle on about the finer details of mulesing (or crutching) sheep. Melbourne city never lost its fascination.

In the early 60s it was a hive of activity, with little laneways crowded with exciting shops of all descriptions. And, of course, we were all single girls on reasonable wages for those days. Looking back I realise that we didn't accumulate as much as we do now. I spent three years in the Army, arriving with one suitcase, and leaving three years later with one and a half suitcases. It seems our money was spent more on living and enjoying life. Going to films or stage plays, parties, parties and more parties; going dancing, eating, sitting in cafes sipping on the new sensation in my life - cappuccino. My room mate, Gloria, and I delighted in donning our casual civilian clothes and walking down to the Myer Music Bowl to watch and listen to free concerts. I don't remember ever having to pay. We'd take a blanket to sit on, and thoroughly enjoyed the experience. During summer we'd don our bathers, grab a transistor radio and beach towels, and go downstairs to the grassed area near the flagpole to sunbathe. Sometimes we'd catch a tram down to St Kilda to swim at the baths, or eat at one of the many cafés there.
Helen McLeod (enlisted 1963)

Figure 34 June Aulfrey and Helen McLeod (right), Southern Cross Hotel, Melbourne, 1964

I loved Melbourne. The cool, often rainy and misty weather was wonderful for someone coming from Perth's hot summers in the days before air-conditioning. I was so happy to get back to a city after almost a year in the bush at Kapooka. Melbourne was so exciting with so much to see and do – lovely old buildings, wonderful restaurants and shops and beautiful public gardens. The social life was great. We were invited to dances, cabarets and balls. There were lots of parties to go to and groups would go to the theatre or the cinema.
Lyn Kane (enlisted 1964)

There were the movies at Kapooka. The films were a bit strange but it was a very good place where you could meet up with the people that you weren't supposed to meet up with, i.e. recruits. You weren't supposed to fraternise but I'm sure that more than one of us met the guys at the movies and had a bit of a cuddle and then disappeared separate ways when the movies finished.
Jenny McLay (enlisted 1962)

Figure 35 Party, Melbourne, c. 1965.
WRAAC members left to right: Cherryl Burrows, Jan Whalan, Hannah Schunck, Maria O'Connell and Irene .....

We had a theatre in the camp area at Moorebank, with movies every week. My room mate took one look at the projectionist and declared that she was going to marry him, which she did. All the Army balls were held there as well as dances every pay week at which the Regimental Sergeant Major (RSM) presided. When National Service, 'the birthday ballot'[12], was brought in, the theatre was converted to accommodation for the troops.
Merle Viles (enlisted 1962)

I remember a group of us would often go to the movies at Graceville – it had the old canvas deckchairs and you could still smoke in the theatre in those days. We'd turn up in our fluffy slippers and hair rollers (covered by a scarf, of course) and some would take knitting. It was like sitting in our lounge room at home!
Kay Owen (enlisted 1964)

One Saturday night Major Crask was on duty and we (girls) were going to a party that night, so I told Thelma Graham (cook) that I'd do the dishes and wash up all the pots and pans and whatnots. Thelma said no other ranks (ORs) and sergeants were coming in for dinner and only Major Crask would be in. I said 'No problem, I'll fix that'. I got the menu board down and wrote: 'Pig's Tits on Toast' and 'Fly-blown Custard' The next thing, poor Major Crask came down, looked at the board and said 'Oooh no Private, I don't think I could have that. No, I definitely can't eat that. Private Farrell you come up the front and I will throw you $10 over the balcony and you can get me a salad from the restaurant next door'. That was it; she didn't have to have Pig's Tits on Toast !!!!! And we went off to our party.
Fran Farrell (enlisted 1963)

I remember my first hangover - my eighteenth birthday. I got so drunk drinking drinks everyone was buying for me. I had NO idea what it would do to me and then had to turn up for work the next morning (no sickies in those days). I had to hose the rug from my room in the barracks as I had been sick and then sit for eight hours with headphones learning Morse code at a high pitch.
Sandy Jarrett (enlisted 1980)

Many a happy hour was spent in the courtyard at 31 WRAAC Barracks where we would either sit and sunbake or occasionally get invigorated and put roller skates on and roll up and down the courtyard. The wet canteen was also a popular haunt. My most memorable night (but not my proudest) in the canteen was tossing my head back, connecting with the wall and promptly sliding onto the floor – out cold.
Lyn Morris (enlisted 1966)

A mixed group of women and men from School of Military Engineering would go to concerts at Moore Park. The best concert I saw there was Alice Cooper's *Welcome to My Nightmare*. You would be checked prior to entry to see if you were bringing in alcohol. Most added the liquor of their choice to bottles of Coca-Cola and brought it in as a soft drink. Marijuana filled the air at these events. I was also a part of a group of WRAAC who would go camping. One of the women at School of Military Engineering (SME) had some camping gear and we, and a group of others from the barracks, would go camping in a national park. One of the women, a female Military Police (MP), was on exchange from England. This was the first time I had heard about Exercise Long Look whereby soldiers would be exchanged between the British and Australian Army for a period of six months.
Christina Morgan (enlisted 1975)

One time when I was off duty at Kapooka I decided to watch the Regimental Drill Instructors (RDIs) training the recruits on the parade ground. One RDI got so frustrated with one of the recruits that he started jumping up and down on his slouch hat! An A Company RDI conducted fix and unfix bayonets training whilst standing on a low tree branch as he thought it safer than being on the ground with the recruits. One of my many recollections of 7 Camp Hospital was that on some evenings the unlocked ambulances were used for extracurricular activities. After the drivers got wind of it the ambulances were locked.
Chris Cain (enlisted 1964)

Further to Chris Cain's story of the ambulances at 7 Camp Hospital. I was rooming with a girl who regularly used that particular amenity. She was quite popular with some of the men who nicknamed her 'Passion Pants'. She came home one night quite flustered and told me that she had been enjoying some 'private time' with a medic in the back of one of the

ambulances, when one of the doctors and a medical orderly rushed out, jumped in the front of the ambulance, and proceeded to race towards the scene of a car accident on the base. PP and her friend had to announce their presence to the two startled men in the front of the ambulance and then proceed to get dressed before having a long walk back to their respective barracks.
Lyn Kane (enlisted 1964)

There was a swimming pool at Puckapunyal that was enjoyed by all. People from married quarters (i.e. soldiers' families) were allowed to use it from time to time but basically it was there for the troops. I went there one day with a couple of girls from the hospital and one of the girls, who had come from Queensland, had this lovely white bathing suit on. It suited her because she had very dark skin and she looked really nice in it until she got it wet. She might as well have not had anything on. She couldn't come out of the pool because it looked like she was naked, so we had to chuck a towel in to her so she could cover herself before she'd get out the pool. After that she only ever wore that bathing suit if she wanted to sunbake.
Marion Quick (enlisted 1965)

I had the opportunity to enjoy my passion of swimming, so I went into training and I represented the Army at the Inter-Service Sports in Melbourne in 1968 and in Richmond NSW in 1969. It was a wonderful life and you were able to excel in many things. I was asked if I would like to make my debut while I was based in Sydney in 1969. Our partners were also Army personnel and were chosen by the Army. I was so happy as I would never have done so otherwise.
Ann Dullaart (enlisted 1966)

Have I mentioned the fun I was having? Life was so carefree and secure. We had good accommodation, good pay, no real responsibilities apart from going to work, and had absolutely great friends. On the completion of my course I was posted to 31 WRAAC Barracks, Melbourne, as the barracks driver. It was a real bonus to meet up with my friend, Maree, again as she was the cook there. Other friendships which were to become lifelong were also started at 31. It was great. While at 31 WRAAC Barracks I was involved in both the Inter-Service Sports and Inter-Service Swimming. Trials for the swimming team were held at

Puckapunyal, not because of its sunny climate I'm sure, and next thing I found myself on the team. At school I had always competed in freestyle and breaststroke. However, my new 'coach' (no names, no pack drill) thought I was a great choice for the vacant spot in the 100 metres backstroke. Since I'm generally up for a try, I set about trying to at least master the stroke. The Inter-Service swimming was held at Olympic Pool in Melbourne, adjacent to the Yarra River. It was packed with loads of enthusiastic, cheering supporters, and off went the gun to start the backstroke. The race was on and the support of the crowd was very encouraging. I was DOING this! I think the first to finish were getting out of the pool when I was making the 50 metre turn. The cheers of their respective Services were amazing. Should I get out at the half-way mark, I wondered. In for a penny, in for a pound, I decided. I was going to get to the end. It was so hard and such a long way and I dreamed of the release offered by drowning! Eventually with the support and encouragement of the Army, the Navy AND the Air Force, I finally managed to drag myself out of the water and took a bow as I got a cheering, standing ovation from the crowd. Note to self – ALWAYS finish what you start.
Deborah Harding (enlisted 1969)

Six of us who worked at 3 BOD travelled together from 31 WRAAC Barracks in St Kilda Road through the city and then to South Kensington in Pat Lempriere's pride and joy – her Holden car. We got to know a policeman who was often on point duty in the afternoons as we made our way home through the city. He would wave and call out hello and occasionally walk over to the car for a few quick words with us. Sometimes when he was really busy he would blow us a kiss as we drove through the intersection where he was stationed. One day when we were waiting in traffic for our turn to cross one of the main intersections in the city one of five American sailors standing on the nearby footpath approached us and very politely asked 'Can any of you ladies tell us how to get to the docks from here?' Pat said we would give them a lift and moved the car out of the traffic so we could figure out how to fit everyone in. Luckily the car was large and had bench seats so two sailors and two girls got in beside Pat and somehow the rest of us squeezed into the back seat with the remaining girls sitting on the sailors' knees. We were a bit worried about the reaction of any policeman we might see but as luck would have it 'our' policeman was on duty and waved us through with a salute and a

big smile. We safely delivered the sailors back to their submarine and Judith Hurst asked the Duty Officer if we could have a tour of the sub. He got a bit flustered and said 'But ma'am there are sailors down there in their underwear!' She said 'We wouldn't mind' and he replied 'But ma'am, the sailors may mind'. We didn't get our tour which was a relief for the rest of us as, unlike Judith, we would have been embarrassed to see sailors in their underwear and we wouldn't have fancied climbing up and down ladders in our straight skirts. We invited the sailors to our mess for a drink that evening and several of them came. They were nice young men, happy to have some female company after several months at sea.
Lyn Kane (enlisted 1964)

Figure 36 Lyn Kane, Cherryl Burrows. Ordnance Corps Dinner. Broadmeadows, Vic., 1965.

Then there were the Merchant Navy sailors. When the 'on duty' staff had to stay on board and got a bit lonely, a phone call would be made to the barracks and I would ask around to see if any of the girls wanted to go out. A taxi would pick us up, we would go to a fully stocked bar and

be looked after and a taxi to get us back to 31 before midnight. The boys that I worked with at 1 Base Printing used to call 31 WRAAC Barracks 'the bat cave' as the girls all 'flew out at night'.
Cherryl Burrows (enlisted 1964)

I invited some of my colleagues from my signals unit over to WRAAC School for a few drinks, which was pretty risky because they were quite devilish. All the invited guests had to sign a book and we were responsible for our guests, so I definitely took a risk doing this. The evening went well and was uneventful, which surprised me. The following morning, I was called up to the orderly room and was shown a framed picture of the Director, Colonel Best, which had been tampered with. My friends had cut out a nude body from a magazine and placed it in the photo frame, leaving her head, complete with the beret, poking out, and the nude cut-out on the bottom. Needless to say, that was the last time my friends got invited to WRAAC School.
Julie Elverd (enlisted 1960)

(In Singapore) we were barracked at Tanglin Barracks, a wonderful place to stay, so close to the local markets and shopping centres in Orchard Road. The food at most barracks was very good. There were so many small markets, beer and milk bars, the NAAFI (British Navy, Army and Air Force Institutes), and an old replica of a British pub called the Crown and Anchor, where the prawn cocktails, crabmeat soup and shark fin soup were to die for. We all thought we would need to book two seats when it came time to fly home. There was always plenty to entertain on days off and after work. We joined netball teams, hockey, football and soccer teams, and played in mixed teams against a lot of sailors off the ships, and some of our boys on Rest and Recreation (R & R). Swimming at the camp pool and regular outings to water skiing were very popular. The boys loved taking us into the swampy areas, amongst the sea snakes, and try to knock us off our skis. These activities helped to keep our weight in check. We had parties at the Armada and NAAFI clubs, and dancing every second night somewhere. We were never in need of partners as we were outnumbered ten to one. We had fun visiting Raffles Hotel to try a Singapore Sling, followed by a trip to Bugis Street to see the sights. On Christmas Eve, a group of us were invited to a cocktail party put on by Cathay Pacific Airlines and held at the grand ballroom at the Hilton Hotel, with marble floors and chandeliers, butter and ice

sculptures made into dolphins. There was an article printed in the Australian *Age* and all three Tasmanian papers with a photo of me in a coolie hat. We had a visit from Queen Elizabeth and Prince Philip. We had practised marching till our legs ached and had car door opening practice too. Got a big wink from Philip when I saluted him as he got out of the British High Commissioner's car.
Libby Flemming (enlisted 1969)

I certainly have some great friendships, particularly from my time at Staff College in 1996. I was one of three women, and the only single one, and there were eighty-seven men. I lived in with three single guys, and there were another nine married people who'd left their families behind whilst they were doing the Staff College year, including some from overseas. We had a Kuwaiti, a guy from Indonesia, one from the Philippines, one from Papua New Guinea. So there was Dee and the Dirty Dozen. We had a great group and on Fridays they used to give me their wallets and say, 'Where are we going this weekend?' And so most weekends I'd set up some sort of a little trip for us to go somewhere. We might go out to Geelong on a Friday night, and go to the movies, we might go to a pub somewhere, or I'd arrange to go to an AFL game or a museum.
Denise Dunn (enlisted 1976)

Close to Kapooka (NSW) was Lake Albert. There was a skiing club that belonged to the Army and we had our own ski boat, so, quite often of a weekend, we would gather on the Friday night, go out and stay there the whole weekend, water skiing and barbecuing and having the odd soft drink. Yes, that was a great pastime. In the winter we participated in things like netball, basketball, hockey, and competed against the local teams and the Air Force. There was always great rivalry between all three - the civvies, the Air Force and the Army. I think we held our own. We also followed the guys with the rugby and on weekends we would also have the guys over to the recreation room for parties. It was a pretty constant social life. Sometimes I wonder how we found time to work.
Jenny McLay (enlisted 1962)

Figure 37 Far left: Cherryl McNamara. Far right: Arthur and Jenny Robertson (nee Jenny McLay). c.1965

In mid-1966, a mutual friend introduced me to Noel, an officer in 7 Battalion RAR (Royal Australian Regiment), which was stationed at Puckapunyal, sixty miles from Melbourne. The battalion was training and preparing for its upcoming twelve-month tour of duty in Vietnam. Noel and several of his friends, other young officers from 7 RAR, managed to get leave to come to Melbourne for a number of weekends before they went overseas. On Saturday nights we usually went out for dinner and then to a dance. On Sunday mornings I would meet them in the city and we would go to the stately old Menzies Hotel for breakfast. I remember walking in the door and my feet sinking into the thick carpet emblazoned with the Menzies tartan. There were beautiful starched tablecloths on the tables and huge white napkins, silver cutlery and flowers. People dining there were very stylishly dressed. I used to feel like a movie star being greeted by the maître d' as we swanned in dressed to the nines on a quiet Sunday morning. The weather always seemed to be cool in Melbourne, so the women would mostly be dressed in winter clothes, and often overcoats. I remember on one occasion that I wore a

cream overcoat with a fake fur collar over a tailored Chanel-type suit. Of course I had matching shoes, gloves and handbag. I was twenty years old and had had over a year to get used to dining out in Melbourne, but I had grown up in Perth and had never been to a restaurant until I joined the Army. That old hotel was one of the most elegant places I had ever been to and I loved it. We would have a leisurely breakfast and then walk through the city hand in hand talking and laughing. It was a very carefree time. Noel invited me to the big Christmas party which was held in September 1966 in the Officers' Mess at Puckapunyal as the boys would be spending Christmas in Vietnam. I drove up from Melbourne with two other girls after work on the night of the party. The mess had been transformed into a jungle setting with camouflage netting draped around, trees in pots and bales of hay to sit on. Everyone was dressed to the nines – the men in mess uniforms and the women in formal dresses. I wore my favourite cocktail dress, a black lace and crepe Susan Small designer original which had cost me two weeks' pay. We had a wonderful time – danced every dance, enjoyed the beautiful supper served by non-commissioned officers (NCOs) who outranked me, not that they knew it, and who called me 'madam' as they circulated with trays of drinks and canapés. Santa arrived and distributed gifts. The women were given ration packs of Cadbury chocolates, dated 1942, which were still edible after all those years. By 4 am I was very tired so asked Noel if we could go. He said we would have to ask the Commanding Officer's permission to leave and when we approached the colonel he said 'Nonsense. You can't leave yet. They are serving breakfast in two hours'. So we had to wait until breakfast was served before making our exit. We walked out of the mess as the sun was rising on a very cold morning with ice on the ground.
Lyn Kane (enlisted 1964)

When I was at Bendigo I bought a motorbike, and loved the freedom, so I would ride as often as possible, but I was also a square dancer so I attended as many dances as I could. In Melbourne you could dance each night of the week and attend at least three dances on the weekend. Quite a good way to keep fit. As part of my duties I was a member of the Mactier Social Club at Watsonia boozer (canteen) and was organiser of social functions for the club. Billy Thorpe & the Aztecs and Normie Rowe were some of the acts that performed there.
Hilary Jones (enlisted 1970)

One February (1956 I think) we were presented to Her Majesty the Queen Mother at Government House Gardens. We had to wear our dress greens. It was so hot that lots of the girls fainted but I managed to stay upright. We also took part in two military tattoos, two consecutive years in Sydney, dressed as Toy Soldiers, which was great fun, especially marching with the massed bands at the finish.
Wilma Stralow (enlisted 1955)

A funny incident happened in the WRAAC recreation room at Kapooka when my future husband told the WRAAC sergeant on duty to get a grip of her knickers when she told him to finish his drink and leave as it was closing time. Fortunately he escaped a charge.
Chris Cain (enlisted 1964)

Security was always an issue with me and I was always challenging what I could do to prove that I was right that their security measures weren't up to scratch. I was a rationed and quartered person at WRAAC School, it was Anzac Day eve and we had been invited by the signals unit to come back after lights out to join them for an Anzac Dawn Service, as they put it, in the mess. They were going to black out the bar because there was a duty officer on who was quite deaf and went to bed very early. My friend and I got transported from Dundas where the signals unit was, back to WRAAC School in Mosman and we put it to a couple of the other girls: 'Are you going to come to the party, because we are?' So we stuffed our beds, put mops in the beds, tied scarves around the mops and had hair rollers lying around the pillow. Our pyjama jackets were stuffed with Zippy Starch bottles looking like arms. We got dolled up and away we went out the gate and nobody questioned us. We got a cab and went to the party. We were there until 6 am, and, when we got back, the duty officer was just putting the flag up. We came in with laddered stockings and mascara running, looking like Wrecks of the Hesperus. She (the duty officer) was very regimental, continued to raise the flag and said, 'Good morning ladies. Been to the Dawn Service, have we?' We just continued on our way, 'Yes ma'am'. We didn't get into trouble for that.
Julie Elverd (enlisted 1960)

I went on a short course to 30 WRAAC Barracks at Mt Martha, Victoria, and we also had some Women's Royal Australian Naval Service (WRANS) who attended. They invited the Army girls to nearby HMAS

Cerberus one night, to have dinner in their mess, and to go to the movies on the base afterwards. We WRAAC and nursing corps girls felt very conspicuous in our green or grey uniforms walking into the cinema full of sailors. Many heads swivelled around and someone called out in a loud voice 'You can relax and enjoy the movie, boys. The Army is here to protect us'. They whistled, clapped, stamped their feet and cheered, and I remember blushing furiously and slinking into the nearest seat just as the lights went out.
Lyn Kane (enlisted 1964)

Figure 38 Character Leadership Course, Mt Martha, Victoria, 1966. Lyn Kane second from left, back row.

My friends and I used to go to pubs and have a drink or two. We didn't get drunk and disorderly. It was more of a social get-together. We went to balls and dances and had lots of parties at this place and that. There was usually something happening. I remember driving to Canberra for the weekend to catch up with an Army friend, Pat Smiley. I was to be driven by Pat's brother, Jack, and three of his mates, with one of the mates

being the driver/owner of the car. I had to wait outside WRAAC School to be picked up by this guy. He was really late and I got a thorough drenching in the rain. He drove me to the Army camp at Moorebank to pick up the other guys and they took me to the drying room until my clothing was dry. We then continued on our journey only to get as far as the outskirts of Campbelltown, before the driver lost control of the car on the wet road. The car rolled three-and-a-half times before coming to rest on its side. Fortunately, none of us were injured in any way. There were four guys and me in the car. We got back to Campbelltown, where the guys phoned for a mate to come and get us. As we stood there on the pavement, late at night, a police car cruised by and stopped. A big, burly sergeant stuck his head out of the window and said to me 'Are you all right, Girlie?' I assured him I was, and they drove off. I spent that night at my sister's home in Bankstown. First thing the next morning the guys turned up with another car and another driver and off we went again to Canberra, as planned. Every time I see Jack (and I do, every year or so), he always says to me, 'Are you all right, Girlie?' and I feel eighteen again.
Marie Taylor (enlisted 1962)

The boys from the Battalion used to drive up the steep slope past our office window and one in particular used to hop out of his truck and put his hand on the accelerator as he walked beside it waving! Sometimes they'd put their feet out the driver's window as they went past. Obviously the truck was engaged in a gear to keep it going but it gave us lots of giggles.
Kay Owen (enlisted 1964)

Thelma, Gloria and I shared a room and were a great threesome. Outgoing, bubbly, happy outrageous Thelma; calm and efficient Gloria; shy me. We laughed a lot in those days. Laughed and talked and slept in late whenever we could on our days off work. I suppose it was all part of growing up. We were all in our late teens and probably needed the extra sleep. I remember Sundays in particular as being the best sleep-in day of the week. No room inspection! Blankets pulled up over my head, I revelled in being able to sleep until lunchtime. We had a record player in our room and would delight in playing Johnny O'Keefe, Elvis Presley and Col Joye songs on vinyl over and over. It was quite on the cards that the three of us would be singing along in loud, raucous voices, sometimes joined in by girls from other rooms. Weekends were always the time to get uniforms clean and ready for the coming week, so the laundry room, clotheslines, drying room and ironing room were all usually occupied non-stop each Saturday and Sunday. I had bad dreams for many years after I left the Army - dreams in which I was going on parade, or to work, and found that I didn't have my uniform ready, or I was wearing a floral blouse instead of my Army shirt and tie, or couldn't find my beret or lanyard.

Gloria would be forever writing letters to her boyfriend Ken who was in the infantry and serving on the Malay-Thai border at the time. They later married and Ken went on to join the Air Force, serving until retirement. I'd always been keen on writing, and I would sit on my bed also writing letters. A fifteen or twenty-page letter from me crammed into an envelope was not unusual for my family back home in WA to receive. Unfortunately none of those letters have survived the years. Most of us girls owned a writing compendium that zipped up around three sides, and contained a lined writing pad, envelopes, stamps, a biro or pen, and, perhaps a small address book. Food was always plentiful and freely available outside of mealtimes at 31. We had access to the pantry area where our cooks would leave bread, butter, condiments, trays of cake or other delicious fattening food. Huge pails of milk were in the fridge and we'd fill up our Birkos and make pure milk coffee back in our rooms to drink with our toast and vegemite or whatever took our fancy.
Helen McLeod (enlisted 1963)

Going home on leave

We were not very popular with other people in Brisbane because it took us five days to get home from Brisbane to Perth and our leave did not start until we hit Perth. We would travel from Brisbane to Sydney on the train - travelled all night so therefore we had a day in Sydney. We'd leave Sydney of a night-time and travel to Melbourne. We'd have another day in Melbourne and, of course, our leave hadn't started yet. We would catch the night train to Adelaide and had four hours in Adelaide before we travelled to Port Pirie, where again we would change to the Trans Australia train for Kalgoorlie. We had an extra ten days added onto our leave because we didn't start our holidays until we got back to WA. In the meantime we had a day each in Sydney, Melbourne, and a few hours in

Adelaide. The girls, regardless of what Service, would get the first-class sleepers and the boys were crossing the Nullarbor in sit-ups.
Christine Negus (enlisted 1963)

LEAVE

The service grants three weeks leave each year, plus public holidays, and provides free travel to your home State.
Local leave is granted after duty hours. After duty the choice of dress is yours—civilian clothes or smart walking out uniform. When in uniform, hairstyles are neat—not bouffant, and the selection and use of cosmetics are included in your training.

I did thirteen first-class trips on the train from Perth to the Eastern States. It was more expensive to fly and days travelling were not considered an issue. I am not really into train travelling these days. It was so boring travelling with the blue rinse set.
Jayne Rickard (enlisted 1969)

I once travelled alone back to Melbourne after being home in Perth on leave. The first night on the train I was put at a table with three older ladies who were very nice. However, the next morning, when I came in to the dining car for breakfast, the steward told me that the Chief Steward had thought I might be happier with some younger company, and he led me to a table which was occupied by three sailors. I did enjoy their company and during the journey they taught me to play poker – not strip poker I might add! On my journeys across the Nullarbor in first-class I would have my own compartment, and at night time, I would leave the blinds up so I could lie in bed and look at the stars which were so bright and looked so close that I felt I could reach up and touch them. One night I fell asleep on top of the bed and woke to find the train had stopped at a small railway station in the middle of the Nullarbor. I opened my eyes and people were standing on the platform very close to the train. Here I was in my blue, baby-doll pyjamas wondering if I should make a move, sit up and pull the sheet over me, but decided that I would never see any of those people again so I just shut my eyes and pretended to be asleep.
Lyn Kane (enlisted 1964)

Travelling home on leave was by train, then plane. Always an enjoyable time. Train travel was exciting as we were always in the company of other Service members, Navy guys mainly.
Andrea Woods (enlisted 1967)

My first leave, not having seen my family for eighteen months and travelling first-class on the train, was simply the best. The trip from Melbourne to Perth took four days, catching the Overland sleeper train to Adelaide, another train to Port Pirie and then on to the Trans Australia train across the Nullarbor to Kalgoorlie. I think from memory it took about twenty-nine hours. On the Trans train the women went first-class, and had lovely single compartments, silver service in the dining room, and an observation car at the end of the train, where the 'oldies' (probably

younger than we are now) could sit and talk, play cards, play the piano, etc. There was also a well-beaten path back to second class where the boys were. We always had fun. A two-hour stop in Kalgoorlie and then the Westland sleeper to Perth. No air-conditioning, only wall fans, and the journey took about fourteen hours – more than double the time it takes now. On one occasion the conductor was convinced that my friend, Lynette, and I had someone in our compartment. What a laugh! Heaven knows where we would have hidden anyone as the compartments were very modest. He insisted on coming in to check and came back again later, knocked on the door and demanded we open it as he again needed to check if there were men in our compartment! We refused, telling him he was a dirty old man who just wanted to 'perv' on us in our night attire. He took the hint and left us alone after that. I have to say the trip back to Melbourne was a bit lacklustre, leaving family and boyfriend for another twelve months, but, once back at 31 it didn't take long for things to return to normal. There were always bouts of homesickness that would necessitate trying to scrounge all the coins to phone home, just to hear Mum and Dad's voice. All too soon the money would be gone.
Cherryl Burrows (enlisted 1964)

Chapter 6

Soldiering On

This chapter aims to give a wider picture of Army life and times for women. Army life in general was a positive experience for the women whose stories are told here. However, in such a male dominated organisation it was inevitable that a number of them experienced discrimination in one form or another. Because female soldiers of the 1950s and 1960s did not do the same training as men, they were not always seen by men as equal and in some cases were treated accordingly. Women who had enlisted to do their bit and 'hold the fort' while men served overseas were sometimes disappointed and angered at this kind of treatment. During the five years I spent in the Army from 1964 to 1969 I found that male soldiers in general were respectful of female soldiers, enjoyed working with us and were protective towards us, but this was not the case with everyone as some of these stories show. Because of the power imbalance, some male officers and Non-Commissioned Officers (NCOs) took advantage of their rank to harass female soldiers who worked with them.

A female officer whose story is not told in this book told of WRAAC Major F who received notification in the late 1970s that she was to receive equal pay with her male counterparts and shouted the whole unit to a drink. The woman relating the story said she was unaware that female officers received less than male officers or that women in general did not receive the same pay as men. She also told an amusing anecdote of this same WRAAC major who was the senior female officer in a unit which received new recruits direct from Kapooka. 'One morning three young lads freshly arrived, saluted Major F and said 'Good morning Sir' as they went by. She quickly pulled them up and cupping her hands underneath her quite substantial chest said, 'See these, these are called Ma'am'. I bet they never forgot. As the passing observer, I never have!' The same former officer told me that she thought 'it was sad that men in the Army didn't know how often they belittled women, that people with ambition were frowned on, that women who weren't married either had to be gay

or frigid and that if women complained they were told they couldn't take a joke'. She also thought that 'it was sad that too many women didn't stay in the Services very long and have careers; they found a husband or partner, got married or lived together, had children and got out.' This former officer enlisted in 1978 at the age of seventeen and served for ten years.
Lyn Kane Dale

I'd recently gone to Watsonia where there was twenty-four hour shiftwork. I was put on with a staff sergeant to learn the ropes and was the only female on shift.[13] Upstairs were the technical staff who were responsible for keeping all the communications circuits we had. At the time when the Vietnam War was on, Vietnam Circuit was Priority One so if there were any garbled messages coming along they were reported straight away. Midnight came and this voice came over the PA system asking, 'Who's doing the supper?' I picked up the phone and said 'I don't know. Who does the supper?' He said 'Well, it will have to be you because you're the only female on shift tonight and the girls usually do the supper'. I said 'Well, I won't be doing the supper. You'll have to send one of the boys down'. I don't think they liked it. However one of the boys came down and I could hear the tins rattling and things getting cooked, then the trolley was coming down the passage and there was coffee, a chocolate cake and everything else. One of the boys had done an excellent job.
Julie Elverd (enlisted 1960)

My first experience out of WRAAC School was when I went to the School of Military Engineering to do the Biological and Chemical Defence Instructors Course. This was 1977 and women were in those days on 75% of the male wage. There was a lot of discrimination. Women weren't really wanted in the military. They didn't have equal opportunity. They were only administrators and assistant administrators in various areas, but things were starting to change. I remember going into the officers' mess, to the bar, and was asking about my room. Because there was no female accommodation the commanding officer had given up his flat in the mess. He was married and lived in the married patch but he'd

given up his flat so that I could have somewhere to stay. Norm the barman said to me 'Here's your key, ma'am'. I said 'Oh, can I have a Bundy and coke please Norm?' He said 'Well I'm sorry, ma'am, you can't because you have to go in the ladies' lounge to order that'. There were three young engineer second lieutenants down the end of the bar sniggering behind their glasses of beer. I said 'Well, I'm a lady but I'm also an officer and this is the officers' mess bar. I would like a drink, please', and he said 'You can't, ma'am, because these are the rules'. I said 'Well, where's the commanding officer? I would like to speak to him'. Norm replied, 'He's in the ladies' lounge, ma'am, with his wife'. I said, 'Would you please present the commanding officer with my compliments and ask him if he would come to the bar'. Just at that moment the door opened and here was the commanding officer coming out to buy his wife a drink. I introduced myself: 'Sir, my name's Lieutenant Denise Dunn. I'm from WRAAC School. I'm here to do this course. I just wanted to order a drink in the bar and the barman told me I had to go to the ladies' lounge'. He said 'Well, yes, those are our rules'. I said 'Well, I don't think it's very fair because I'm an officer'. At that moment his wife appeared behind him and she was very direct. Her name was Beryl and Beryl said, 'Alan, what's going on?' He replied, 'Oh, this is Lt Dunn'. She asked 'What's the problem?' I said 'Well, I was just trying to buy a drink in the bar but I'm told I have to go to the ladies' lounge'. Well Beryl, bless her soul, said 'Alan, that's a ridiculous rule, change it this minute, buy the girl a drink and come back out in the lounge'. So that's exactly what happened. The rules were changed and I got my first drink in the officers' mess bar.
Denise Dunn (enlisted 1976)

Women were not required to do duties (at Singleton, NSW). I helped out in the mess kitchen for a few days with scraping plates and putting clean ones away but when Madam WRAAC found out this stopped. There was much discussion about women not doing duties, although it was expected you would babysit when there were functions on in the messes. I helped out as a wine steward at the sergeants' mess for some of the formal dining-in nights.
Christina Morgan (enlisted 1975)

Figure 39 Denise Dunn abseiling

When I enlisted in May 1973, female soldiers were not permitted to use military weapons, but by the time I was posted to 4 Training Group, this had changed. Although I had not received any weapons training, I was instructed by my Commanding Officer (CO) to attend a shoot at the rifle range. I was handed a self-loading rifle (SLR) and told to lie down in the dirt and shoot. I froze, so Major K lay down with me and ordered me to shoot. After many rounds, I received significant bruising on my inner arm. I did not know how to hold the rifle, so it kicked back far too much. I was then given a submachine gun and told to stand in line with other staff and shoot. What the instructor failed to inform me was that once I touched the trigger I should release it. I did not know this, so, of course, the submachine gun (SMG) kept firing and everyone laughed. By the end of the day, I had a severe headache and nausea and my orderly room colleague, Sergeant Jenny Saunders, drove me to the toilet block where I vomited for some time. It was obviously stress-related.
Maria Adamczak (enlisted 1973)

I first fired an SLR on a range in Singleton when the unit went out for a shoot. I was asked if I wanted to participate and said yes. I remember being surrounded by about four men when I first fired as there was concern I might panic. I was told to put the rifle in the soft part of my shoulder but did not find the right spot that day or at any other time during my service. The firing didn't worry me but the kick back was extremely painful. To participate in the event, I was issued with my first set of greens as well as black boots and gaiters. The boots were ankle height and the gaiters sat over the top of them and added height to the boots. The boots had to be polished and the greens ironed.
Christina Morgan (enlisted 1975)

I saw a lot of discrimination in regard to equal pay for women, women in the bush, and particularly 'I am not taking orders from her'. The equal pay saw many of the senior other ranks (OR) men show that the feminism line was not going to work. During my time as head of a shift I had to experience many young male soldiers who were defiant of me giving them orders. Sorry guys, I have earned my stripes and you will do it regardless. What made me leave the Army was a male sergeant who kept discriminating against a young male soldier who then attempted suicide. Even after that the sergeant told him he was weak. I went in to bat for the young private and had lots of battles with the sergeant, so much so, that I walked out of Central Army Records Office (CARO) one afternoon and booked a trip to Europe. When I returned (to work) the sergeant said to me 'Are you going to settle down and do what you are told?' I said 'Nope, I am applying for discharge', walked to the Orderly Room and started the process. One main sadness was that I could not lose weight and run like many of the girls. I joined the Army to be the back defence for the guys who served overseas. They even put me in hospital and I did not lose any weight, but I felt that it was a discrimination against me as a person. I always did my job to the best of my ability, and I certainly did play sport. Those days were cruel to all female soldiers if they didn't have the fitness. I was sick and tired of trying to do the physical training (PT), run five kilometres and the whole concept of a male-driven unit.
Denise Darbey (enlisted 1973)

What shocked me was the calibre of female officers, their lack of manners, management (skills) and resentment of someone who was raised in the heavily disciplined environment of boarding school. There

I learnt the protocols which ended up giving me some public relations roles at our unit at times. There was nothing particularly special about me but I did know how to talk to the Governor-General's wife and could sit down with a street kid in the gutter and have something to talk about. That comes of always being the new kid in school.
Joie Norton-Willson Scheinpflug (enlisted 1967)

The small amount of discrimination that did occur originated from the WRAAC officers: we weren't allowed to stay with the other male course members during driver training (what DID they think we might do covered up in those unglamorous overalls and berets?) and certainly one or two WRAAC officers made my life a little difficult when they discovered that I was on good terms socially with a couple of male officers – and then had the cheek to marry one! The speed with which my discharge papers were assembled was astonishing once news of my impending marriage came to notice.
Kim Lake (enlisted 1966)

During recruit school I'd put on quite a bit of weight because the food was exceptionally nice there and we had three meals a day even though we were on the parade ground quite a bit, but anyway the weight went on. I was wearing a pair of slacks to a sports day at Balmoral Beach. When I arrived the CO called me up and told me that I wasn't to wear the slacks again - that only people with ballerina figures should wear slacks and that wasn't me.
Julie Elverd (enlisted 1960)

Discrimination - that's a big one. I guess (that) we were paid as nurses and not as officers? Who knows? I accepted things as they were, I guess, and chose to work around them.
Ann Hall (enlisted 1967)

I was transferred from Adelaide to Melbourne, in 1974 as Staff Officer Personnel Services. I was still a captain and was never promoted. In the Australian Regular Army (ARA), when a female soldier wanted to marry, she would have to leave the Service but by the end of 1969, permission was granted to marry and remain in full-time service. However married women paid the penalty of forgoing further promotions. I was a captain for many years, when I was qualified for major long before I left. That

was the penalty we paid for staying in the Army. I felt particularly sorry for the other lasses I knew. It was far worse for corporals expecting to be sergeants, but never to be promoted, than it was for me to accept. Married Servicewomen married to Servicemen posed a difficulty for Army Headquarters (AHQ) when it came to postings.
Iris Petrass (enlisted 1961)

When I was quartered at 10 WRAAC Barracks, Brisbane, we used to make a sandwich and take a piece of fruit from the barracks to work for our lunch. We were supposed to be supplied with all meals. Suddenly the cook was told not to give us any food as that was not a requirement by the WRAAC Barracks. We asked why we had to pay for our lunches and were told to use the men's mess near where we worked. This turned into quite an ordeal for the four of us. We had to walk up the hill past the male quarters and men who delighted in running out from their showers in towels and yelling at us. When we arrived at the mess, we were made very unwelcome and told to sit at one end facing the wall as the guy with the porn magazines came in to make his sales at the other end and he liked to hold up his stock and show it around. Needless to say, we didn't last long and ended up buying our lunch every day at the shopping centre.
Kay Owen (enlisted 1964)

Margaret Callow was, to my knowledge, the only Aboriginal recruit on our course. It was during this time that I witnessed racial discrimination for the first time. Margaret was my closest friend and during weekends we used to do the normal things, such as movies and sightseeing. When we lined up to be served at local eateries (in Canberra), Margaret was totally ignored by the retail assistants. This happened too many times. There were other examples that upset me, when civilians who lived at Gowrie would ask very racist questions such as 'I bet you have lots of brothers and sisters?' For the first time, I felt I understood discrimination that went far beyond the 'new Australian' tag I was often given when young.
Maria Adamczak (enlisted 1973)

I was at 4 Signal Corps at the time and was offered promotion to sergeant but decided not to accept as I would not be able to take my husband to any mess functions as he was a corporal. The same thing did not apply to male sergeants with a corporal wife. Barracks duties were

another area of contention. When my husband went in to do a barracks duty he didn't have to pay Rationed and Quartered (R&Q) but I did. Do a duty and pay for the privilege!
Hilary Jones (enlisted 1970)

In 1975 I was attached to the State Government Welfare Department and completed the last of their In-Service Welfare Officer courses. We did our theory at the Institute of Technology (now University of South Australia) and it all formed the first year of the Associate Diploma in Social Work. I realised how closeted and institutionalised I had become and just loved being a very well paid sergeant Level 6 (highest increment) student. I loved university campus life, the freedom of dress and intellectual discussion. I had not disclosed to many that I was in the Army. I naively agreed to a photo shoot for recruiting purposes and did not realise it was for Army Reserve recruits. One morning as I was dashing through Adelaide University and University of SA from my car park behind Government House, I was dismayed to discover my face on a recruiting poster for Adelaide University Regiment. My cover was blown and as I walked into a large lecture theatre of eighty-plus people the lecturer drew attention to the poster and me!

At university, I chose to write about conditions of service for Army families for a sociology assignment:

- Families being labelled 'administrative liabilities';
- No consideration given regarding children's education;
- Timing of postings, e.g., mid-year;
- Lack of consideration for spouse careers;
- Lack of consideration for any special needs family members may have;
- Lack of support for wives and young children when husbands were away;
- Low socio-economic trust housing in the very worst suburbs.

I was encouraged by my lecturer to submit the paper to the Army so I sent it to a former Psychology Corps captain that I had worked with in Perth who had risen in rank to full colonel and headed up Personnel Branch in Canberra. He was very encouraging and respectful of it and

obviously passed it around. The outcome was my husband being contacted by the head of Infantry and being told to tell his wife to pull her head in, as she would impact on his career. At the end of the next year, my husband was posted to Melbourne and after all the financial investment (by the Army) in my social work training I was not offered a posting in Welfare. I was pregnant and chose to decline the offer of working in a Pay Office. Instead I went on early maternity leave. Ironically some eight years later, in 1983, when we returned to Adelaide from an interstate posting, I was an Army officer's wife with three young children and I was invited to be the civilian rep in the SA group responding to a Commission of Enquiry looking into conditions of service for Army families. Almost every recommendation I had made all those years earlier was addressed and implemented.
Jayne Rickard (enlisted 1969)

I wrote a letter to the local newspaper about a lost child, who people in the district had gone out to search for, and my husband was told by his commanding officer to tell me never to do that again. During a formal dining-in night in the mess when I was seven months pregnant I was told by my husband that I was not permitted to go to the toilet as protocol dictated that no-one was allowed to leave the table until the mess president rose to leave the table.
Trish Courtney (enlisted 1962)

It was while I was in Canberra that I experienced the greatest level of sexual abuse in my life. I will not divulge the rank or position of the individual, except to say that it was not my immediate supervisor. My reaction to this physical abuse was to strategically avoid the risk, but all I could manage was to minimise it. This abuse would be referred to as sexual assault in current terminology as it was physical. I did not discuss it with my husband or anyone else. I just ignored it. Needless to say that when my husband received notice of posting to Sydney, I was very glad to leave Canberra.
Maria Adamczak (enlisted 1973)

Discrimination? Yes, but most of the time we dealt with it. It didn't invade our every moment but I was referred to as 'Germaine' a few times. My mother was a doctor during World War II so our family was very independent in all roles. I was Norton-Willson from 1967 to 1969 and

then Scheinpflug from 1969 to 1974 so most of my service was as a married soldier. The thing is (that) there were people who resented us. One bloke thought it appalling that my husband was getting Marriage Allowance for me. He could not see that his wife working in Myers was in fact the same as my working. I really capped the argument by saying 'Well, in an equal world I should be getting Marriage Allowance for my husband as well!' Needless to say, subject terminated. The Officer Commanding (OC), who knew my husband well, had a flea in his ear. He made me do a Living Out Application every month and changed our roster so my husband was going to work when I was coming home. We were so professional about our jobs. You would never have known we shared the same bed. I outsmarted the OC by applying for the day job as Non-Commissioned Officer (NCO) in charge of Enoggera Barracks COMMCEN (Communications Centre). At least we knew where I would be most of the time. (In the Army) we were used to solving a problem and working to fix it. I had loads of confidence as I was used to criticism. My mother was a master at bringing you down to a working level.
Joie Norton-Willson Scheinpflug (enlisted 1967)

I enjoyed every minute of my five years at 1st Training Group. I worked with many wonderful people. My only memory of discrimination was very early at the unit when I was told that it was my job to make the coffees. Now I may only stand at 151cm but this didn't stop me replying, 'I didn't join the Army to make coffee'. That was the last time I was treated in this manner.
Annette Wilson (enlisted 1977)

Everyone in the unit except our civilian typist went to Northam Army Camp, sixty miles from Perth, twice a year for two weeks so the cadets could do their training. The conditions were very primitive. We worked in unlined fibro huts with corrugated iron roofs, boiling hot in the summer and freezing in the winter. We worked fourteen days straight while we were in camp and long hours – sometimes twelve to fourteen hours a day just to get all the work done. I was also ordering the rations for the camp so that was a priority, keeping everyone fed. The NCOs were mostly warrant officers and sergeants who were drill instructors for the cadets, so the only corporals (several males and me) ate in the sergeants' mess. The camp was situated near the main road, but in the bush several miles from the town. All power was supplied by generators which were turned off at 10 pm and then the camp was in total darkness. I was the only woman in the camp and was quartered at the old hospital, a large fibro and iron building dating from before World War II. I was nervous being there on my own at night as there were no inside locks on the doors. I think from memory there were padlocks on the outside which could secure the building when it was unoccupied. One night I was in bed in my room after the lights had gone off and I heard a car pull up out the front of the building. I looked out the window and in the moonlight could see it was the CO and he was drunk as he staggered to the front door. I quickly went into the main part of the hospital and hid under a bed. He came into the building, called out my name and said not to be afraid – he had just come to kiss me goodnight. Forget about being afraid - I was terrified! He blundered around the place looking for me and then gave up and staggered out. The next evening about ten blokes and I were in the mess when the CO came in and said in quite a loud voice 'I've been thinking we need to find a bloke for Lyn.' I had met my future husband several months previously but did not tell him this. He said, 'What is wrong with these blokes here?' I said, 'Well sir, they are either married or they're too old.' He said, 'I'm not married.' I thought about it for a couple of seconds before I replied, 'But you are too old.' He was fifty-three and I was twenty-two. I realised I was shooting myself in the foot by saying this but thought it would probably discourage him from harassing me in future. It worked. He never spoke to me again unless it was barking out orders to me regarding work. He was pleasant to everyone else but, of course, I had hurt his pride and he never got over it. One day he marched through the Orderly Room with a visiting officer and said 'Two coffees in my office, Lyn. NOW.' Reluctantly I went out to the tea room grumbling to myself that it was always me who had to make the coffee – not one of the men who were the same rank as me. I discovered that our civilian typist had given the leftover milk to the stray cat that lived at the barracks. I ran outside and whipped the bowl of milk out from under its nose and poured it into the coffees. The visiting officer later said 'Thank you, corporal. That was a lovely cup of coffee.' I said 'I'm glad you enjoyed it, sir.'

Lyn Kane (enlisted 1964)

"Of course, you realise that this is most irregular, Corporal . . . you're supposed to go through channels before you get to me."

After recruit training, the next available Clerical Administrative course was being held at Randwick Barracks, NSW, but there was a gap of several weeks. WRAAC School transferred me into staff barracks as a sergeant's mess stewardess. I cooked weekend breakfasts and did all the other associated waitressing and cleaning duties. One incident that I do recall whilst on 'staff' waiting for trade training was when my room mate had put her heater into the plug on my side of the barracks, and gone to sleep leaving it on. As I had gone to sleep before she had, obviously at the time of room check by the Duty Officer, the heater was still on. In those days I was a solid sleeper, not easy to wake, so the Duty Officer, a captain, could not wake me using verbal cues, so she hit me across the face, leaving me with a black eye for several days. When I tried to explain it was not my heater I was informed that it did not matter, it was my power point, my responsibility. As all soldiers know, ignorance is no excuse. This captain was not a favourite of mine and thankfully was employed in a support posting role, not an instructional one where I had to see her in the classroom or on parade ground.
Maria Adamczak (enlisted 1973)

Shortly before I joined the Army in 1965 I had my hair dyed jet black. It was done professionally and looked good. Anyway after being posted to 3 Camp Hospital I went down to Melbourne and had my hair changed back to my natural colour. Once again this was done professionally. The next afternoon, before I was due to go on duty Matron Potter read me the riot act about changing the colour of my hair without permission. She went on to say that my papers would need to be changed as my hair colour was now different to when I first enlisted. Needless to say for the rest of my service career I didn't again change my hair colour. Sometime later I was sent to the Army School of Health at Healesville, a picturesque little town not far from Melbourne. The weather was starting to warm up and they had a lovely little swimming pool there, so I thought I would go for a swim. My bathers were a two-piece bathing suit. They weren't bikinis; they were two-piece - just like a one piece cut in half so there was basically no skin showing. I'd only been in the pool a few minutes and I was summoned to the matron's office and told in no uncertain terms unless I had a one piece bathing suit I wasn't to go back into the pool so for the rest of the course when it was hot all I could do was sit on the side of the pool and dangle my feet.
Marion Quick (enlisted 1965)

I started my driving course at Broadmeadows, Melbourne, at the end of July. There were four girls and approximately twenty-eight males (that ratio thing again). We lived at 6 Signals Regiment, Watsonia, and were bussed to Broadmeadows every day for the two months of the course. A great memory was marching to morning tea, the warmth of the little shop, pasties with tomato sauce and Creedence Clearwater Revival's 'Proud Mary' on the jukebox. Magic! We learnt mechanical principles, which we were tested on early in the piece and had to pass in order to stay on the course. Fault finding, wading and fording and towing were all part of the course. During the wading and fording, which was carried out in Moonee Ponds Creek in the middle of a Melbourne winter, we had to wade through the water and then guide the vehicles through. On that particular day there was a massive PR campaign underway. Channel 9 was there, also the Melbourne papers and the Army PR machine. We were freezing and caked in mud as we had to pose for more and more 'action' photos. At one point when we were supposedly pushing a 'stuck' Land Rover up the creek bank in the mud, the driver very slowly moved forward as more cameras clicked away. I was sinking into the mud and, afraid I was about to fall into it face first, pushed myself off the vehicle only to lose my balance and land right on my backside in the dirty mess. Now I was wet, very muddy and cold! It was comical to see myself on the news that night gently dipping my backside into the creek to try to clear the build-up of mud. Many years after when I was in the Army Reserve the Quartermaster came to me and said 'Is this you?' Sure enough it was a terrific full page photo of the four of us girls looking very muddy atop the bonnet of a Land Rover having a welcome, hot cup of coffee. The picture was featured in a *Triad* magazine with an article about forty years of women battling discrimination in the Army.
Deborah Harding (enlisted 1969)

One course I attended in 1964 consisted of about twenty males and five WRAAC and it was held in an all-male area at Broadmeadows. We had no problems whatsoever. For the three years I was in the Army the areas I worked in had more male soldiers than female soldiers. In my experience WRAAC were treated with a great deal of respect by all other military personnel. I cannot think of one incidence where I felt that I was being treated with anything other than respect or caring.
Helen McLeod (enlisted 1963)

Figure 40 Deborah Harding and fellow WRAAC drivers, 1969

During my trade training at the School of Survey at Bonegilla in 1970, I was tasked to go to the School of Catering which was in the same area. On knocking on an outside door to one of the huts it was opened by a guy in a white uniform with a white apron over it. His greeting was, 'What are you?' to which I replied, 'A female soldier'. 'I don't believe you', he said, 'where do you come from?' I told him and then gave him the message I was to pass on. At this stage there were only three WRAAC in the whole of the Albury-Wodonga area and I was one of them.
Hilary Jones (enlisted 1970)

I recall having a verbal run-in with our Regimental Sergeant Major (RSM). He was using profanities within my hearing and I reminded him that I was close by. He replied rather sternly that if I was a lady I would not know what he was talking about.
Andrea Woods (enlisted 1967)

While living in the Officers' Mess at Maritime School we had to share the laundry with the officers on course. I and another girl went to do some washing to find two Papua New Guinea (PNG) officers there. They were talking amongst themselves and discussing us in great detail. What they didn't know was I had spent seven years as a child in PNG so understood everything that had been said. I told them that they should not be so rude about women soldiers and that what they had said was unacceptable. If we could see them blushing I dare say their faces would have been crimson. Every morning after that I was greeted with 'Good morning Lik Lik missus' which translates to good morning little Mam.
Catherine Sanders (enlisted 1982)

I was asked by a warrant officer how he should refer to a group of female soldiers, to which I replied, 'They are ladies so call them lady soldiers'. Then it was an appropriate term. Lady soldiers were a confirmed part of the military. On some fronts a woman should always be a lady. After I was promoted to sergeant with this subsequent entree to mess life a senior WRAAC officer quietly said to me 'A lady always knows when it's time to leave the mess'.
Bethmay Watson (enlisted 1975)

While at 6 Sig our female officer (arranged) for the females to take a bush trip to Lake Eppalock near Bendigo. We scrounged our gear from

the guys and were trucked out to the site, there to learn how to erect 11 x 11 tents, light fires, light kero lanterns, cook and set up latrines. The next day we were sent out in groups with radios and maps to perform tasks allotted to us. The only male with us was the School Sergeant Major (SSM) and he was our instructor for the camp.
Hilary Jones (enlisted 1970)

While on the Sig (Signals) desk at Training Command lots of things passed me. After long weekends it could be quite harrowing as we were often the first to get NOTICAS sigs (Notification of Casualty), usually personnel who had tried to get home and back in a very short time and had taken risks driving, etc. Very sad and often I found myself trying not to cry. After one very big weekend and little sleep I collected the usual envelopes and opened them. This was in the days of 'Kangaroo' exercises.[14] In my befuddled state I thought we were really at war. After several coffees, cigarettes and rereading I came to the realisation of what I was reading. Goes to show what jumping to conclusions does.
Catherine Sanders (enlisted 1982)

in the WRAAC you learn to stand on your own feet

You come into contact with hundreds of different people with varying points of view. This teaches you tolerance, self-reliance, self-control, and judgement. You gain experience, to enrich your understanding of the daily business of living. You develop interest and companionship. You have the chance to express your personality to the full. You appreciate a rich, full life.

The WRAAC is honoured to have Her Royal Highness Princess Margaret as its Colonel-in-Chief.

I remember on day one the commandant saying, 'Being at Staff College is a bit like having a three-legged stool. One leg is about work, one leg is about family, and one leg is social. So to be here and do a really good job you have to make sure that all three of those legs are balanced because if one of them is longer than the others you will fall over'.
Denise Dunn (enlisted 1976)

I didn't have any military background but I think I adjusted very well to Army life. I enjoyed living in barracks accommodation. I loved all the marching, the music, the discipline, meeting all kinds of people and everything about it. All in all I think we had a great time. Even though I was probably shy and reserved I don't remember having any problems.
Julie Ross (enlisted 1967)

During my recruit days, I met up with a fellow soldier, Nancy Powell, who has been my best friend, sister and a wall of support to me. We have been best mates for forty years. We have gone through pretty bad times; good times and this past year celebrated our friendship by going to New Zealand as a special trip. Nancy has served forty years in the Army and now she is in her retirement year. Both our husbands know we are the 'Queens of the Chatter Boxes'.
Denise Darbey (enlisted 1973)

I worked at Central Army Records Office at Albert Park towards the end of my three years of service and I loved working there too. There were many very high ranking military people working there also. I remember one day walking along the passageway and a voice boomed out at me from the other end, 'Straighten up there, soldier!' I was really shocked that the voice was directed at me.
Helen McLeod (enlisted 1963)

I was posted to 6 Signal Regiment, Watsonia. I concluded my trade training on the Friday of the Australia Day weekend and was forced to travel from Albury to Melbourne on that day. Arriving in Melbourne on the train I discovered the Regimental Travel Office (RTO) was not manned, so I had to find my own way to Watsonia by train.[15] This involved an underground hike, carting my suitcases from Spencer Street Station to Flinders Street Station, boarding the next train to Watsonia, and hoping that there would be a taxi at least to take me to the barracks.

No such luck! I asked the attendant where I had to go and how far. 'That way and just up the hill. You can't miss it', was his answer. I can remember being very intimidated by the fact I didn't know just where I was going and what I would find. I set off and it got darker and darker the further I went from the station, with open paddocks on one side of the road and a few houses on the other. I kept walking and after about a mile or so I finally reached the front gate which was manned by Military Police (MPs). I was exhausted by this time and was lucky enough to get a lift with the duty wheels to the WRAAC Barracks, only to discover they were not expecting me until Tuesday and there was no room allocated for me. Welcome to Watsonia!
Hilary Jones (enlisted 1970)

Figure 41 Hilary Jones, 6 Signals Regiment, Watsonia, Victoria, c.1970

Not long before my discharge the company that delivered the pays went on strike. I and the other clerk in the orderly room were 'asked' to accompany the Pay Sergeant to go and collect the pays in the unit station wagon. We were both given 9mm pistols. As you do, both of us checked our weapons. No rounds. 'OK, when do we get them?' 'You don't!! If anybody comes that close throw your pistols at them!!!!' Needless to say we were both totally gobsmacked. The pays were collected without incident. My husband was furious when I told him.
Catherine Sanders (enlisted 1982)

My only time in the bush was on my Subject 2 course. We headed into the Shoalwater (Qld) area with a young, inexperienced driver and the padre[16] in the front of the Land Rover, while I sat on a photocopier in the back. We arrived, set-up and bunked under a big truck. We managed to keep the computer going until the second last day, and it proved to many that the instrument could be a good thing for the Defence Force. The CO decided to fly me home to Amberley instead of me waiting the extra days to travel home. I was taken to Rockhampton Airport and as the plane arrived it blew a tyre. We had to wait nearly eight hours for another tyre to be flown up and replaced while we sat on the tarmac with little food or water. I finally got home and Mum said to me, 'Daughter, you absolutely smell putrefied like a dead mouse'. I had to wash my hair five times, and bath twice that night. I could still smell the remains for a week.
Denise Darbey (enlisted 1973)

I was required to go out to Dundas every week to attend a CO's meeting. The CO (Commanding Officer) was away and another officer was acting in his position. The OC's (Officer Commanding) son was on holiday from university and was employed as a groundsman for a short period of time and would drive his father to work every day, drop his clothes and get changed in the car park. The staff at Dundas were waiting for me to arrive and they said, 'Quick, come and have a look at this'. We were two storeys up, looking down on the car park and in drove the OC in his Mercedes Benz driven by the son. The father got out and came to work and the son got out, opened the boot, dropped everything and put his overalls on. In the next block was the Lottie Stewart Hospital where a lot of nurses worked. I think I was a warrant officer at the time, and I said to them that that was disgusting and the female staff didn't have to look at that sort of thing and neither did the nurses. So I was egged on and, of

course, I was up for anything, and they said, 'Well, do something about it'. The RSM, who was a bit of a card even though regimental, was encouraging me to do this so I went into his office, locked the door, put a piece of rag over the telephone, rang the OC and, pretending I was the matron from the Lottie Stewart Hospital, I said that I wanted to complain about a young lad who was dropping his daks and doing it beautifully, but my nurses didn't need to have to put up with it daily, that we'd had enough of it, and could he put a stop to it. Of course everyone was in on the joke and we were just laughing and laughing in the corridors. That was about two weeks before I was discharged and everyone was under threat not to mention it so I got away with that one.
Julie Elverd (enlisted 1960)

I loved dancing and before I joined the Army I used to go to dances two or three times a week in Perth with friends. I heard that there was a dance on Saturday nights in Wagga Wagga and one of the other girls and I decided to go. We arranged for the duty vehicle to collect us and the driver dropped us off and told us the time that the duty vehicle would be leaving for the return trip. We went in to the dance and there was a fairly large crowd there. We sat down and waited for someone to come and ask us to dance. No one did. The locals just totally ignored us. A couple of airmen (in uniform) from the RAAF base nearby asked us to dance. None of the local girls would dance with them either. The locals did not want to have anything to do with the Army or the Air Force though were quite happy for us to spend our money in the shops in town. One evening the Army was performing a Beating the Retreat ceremony in Wagga and a few of the girls from the barracks decided to go in to watch it. We got a lift in the bus with the band members and soldiers who were taking part in the parade but the women had been told to wear civilian clothes and not uniform. After the parade three of us girls were walking down the footpath towards the bus when four young local lads started walking side by side towards us taking up the entire footpath. We thought when they got to us that several of them would move to one side and let us pass but they kept walking and forced us off the path. Being used to men being polite to women we were stunned and one of the girls said, 'Why did you do that?' One of them replied, 'If we thought you were ladies we would have allowed you to walk on the path but you are all just sluts'.
Lyn Kane (enlisted 1964)

As I (had) enlisted in Sydney, I decided to change my Next of Kin (NOK) to my brother who lived in Mt Isa, Qld, to get a free trip one year. It worked, but because the trains out of Mt Isa were only twice a week I finished up being twenty-four hours AWOL (absent without leave). The CO was not happy!! Neither was I as I lost a day's pay!
Merle Viles (enlisted 1962)

When Winston Churchill died there was a huge commemorative march held in Melbourne. We formed up[17] in the area of Boathouse Drive on the shrine side of the river and the formation consisted of Army, Navy and Air Force. We marched into the city, past Parliament House, around the city and back to Boathouse Drive. It was a long march on a very warm day. I remember our Officer Commanding saying to me after we'd arrived back at Boathouse Drive, 'I'm surprised that you lasted the distance McLeod. Lots of soldiers larger than you didn't'.
Helen McLeod (enlisted 1963)

Figure 42 Sir Winston Churchill Memorial March, Melbourne 1964

When Sir Winston Churchill passed away, Melbourne put on a Memorial Parade. Girls were bussed in from everywhere, even interstate. What I remember about this parade was that the only time we were in step was when we were doing the salute at the Town Hall. We had a brass band in front of us and a pipe band behind us and they were playing different tunes! They only stopped playing as we passed the Town Hall as there was another band there. After that we skipped our way down St Kilda Road. The front of the squad was in step with the brass band and the back of the squad were in step with the pipe band. We had a male RSM marching beside us and he was yelling at us to keep in step.
Jean Fullerton (enlisted 1963)

Doing General Duties (GDs) at 7 Camp Hospital, Kapooka, we worked shiftwork and once after I had just finished my two days off and was on my first day back at work I started to feel feverish and went on Sick Parade. Unless personnel were really sick, anyone who went on Sick Parade was given two disprin and sent back to work. The matron came out of her office and saw me sitting there – a lone female amongst about fifteen male recruits. She asked me about my symptoms, took my temperature and announced that I had rubella (German measles), but said I had better see the doctor anyway. He duly arrived, looked down the line of patients and asked the matron 'What's wrong with her?' She said 'She has rubella'. He asked 'Is she pregnant?' The matron said 'I don't think so but I can ask her'. Several of the male patients leaned forward to look me up and down. The doctor said, 'It's better that she has it now than later when she is pregnant. Give her two days off and tell her to go to bed. She doesn't look too good'. Matron said, 'But she has just had her days off for this week'. He replied, 'Unfortunately you are going to have to give her another two'. I went back to my room, collapsed into bed and slept for the next nine hours until my room mate woke me to give me a meal that she had brought up for me from the kitchen.
Lyn Kane (enlisted 1964)

They'd caught some girls who were AWOL from WRAAC School and they were billeted at 31 WRAAC, in the Regimental Aid Post (RAP) or sick room. These girls had to come up every hour to sign the book. The first night was OK and so was the next day. I was on Phone Duty the second night and when it came to 9 o'clock and they hadn't turned up I told the sergeant on duty, Margie Scanlon, 'Your girlfriends haven't

signed in'. She ran over in a huff and a puff and they'd gone, but they left a lovely note for Captain Reynolds, thanking her for her friendly hospitality.
Fran Farrell (enlisted 1963)

Maritime School and Training Command shared the same RAP. I had a phone call from my, then, very new husband saying he was very confused. He had been told to report to a doctor in the city and had no idea why, and when he rang the RAP to ask what it was for he had been told it was an Ob-Gyn. They had got us mixed up!
Catherine Sanders (enlisted 1982)

I went to Canungra to do my Subject One for Warrant Officer. I hadn't even started the course and I had to come home because my young son, Andrew, died of croup. They flew me back. The unit I was leaving and the unit that I was going to were very supportive. We proceeded on our posting to Brisbane because I couldn't stay around family here in Victoria. We still had to get on with life. Even though my husband was great and work was good I didn't want to live. I was in mourning and was basically existing day by day. I was so glad that we had the military family that gave us that extra support. I've been blessed during my Army career. When they said that Pay Corps could give a guernsey for deployment in Somalia I said, 'Can I go?' I think at that time they were trying to give females in the Army more of a profile so that Defence was seen as employing females in different roles. When we became Pay Corps in the July of 1990 I was the first female Warrant Officer in the corps. I got chosen to go to Somalia and then they had to choose another female to go to the same unit so that I had company. Apparently they had chosen ten male MPs, so one of the males was taken off and replaced with a female. I was still in that frame of mind that they can shoot me if they want - I was not really all that interested. I missed the family, I wanted to come home but I wasn't anxious about being overseas either. It didn't worry me. I poured myself into my job. I was in Somalia from January to May of 1993. It was active service at that particular deployment. I was working in the headquarters at Mogadishu, in the old American Embassy that had been converted to accommodate all the army and coalition forces that were there. We were behind a big brick wall and you only ventured out when you had others with you. Once a fortnight we would conduct Pay Parade at Baidoa where 1 RAR (Royal Australian Regiment)

were doing the humanitarian stuff. One day I was part of the security detail going out of Mogadishu to the airport to pick up some people from Baidoa so that they could get away from the desert to give them a change of scenery. One of those that we picked up (was) a nurse who had worked at the Puckapunyal Hospital where my husband had taken Andrew the night that he passed and this lieutenant had actually worked on Andrew. Of all places I met him there. We had a big chat and I asked him questions that I couldn't ask my husband because I wasn't home when it happened. I got my answers over there and it was like a massive load had been lifted off me. I was (then) looking forward to coming home and keeping my head down so that nothing happened. I later had another child, Bradley, who was born in 1997.
Anna Celi (enlisted 1976)

Like when Elvis died or when man landed on the moon, everybody remembers where they were on 9/11. I was Director Protocol and Visits in the Department of Defence and I was in Darwin with forty-three overseas defence advisors and attachés and their spouses on our annual tour of Australian Defence Force (ADF) capability. We had been to Melbourne and Adelaide, and we were going to finish our tour visiting Uluru after Darwin, which would have been the highlight of the attaches' and advisers' time in Australia. I had a Hercules Aircraft and we were on the second last leg of our nine-day trip. We were sitting on the wharf at Darwin at a restaurant right beside a US Marine Expeditionary Force ship and planning on having this magnificent seafood dinner and then having a look at the ship after dinner. During the first course we suddenly heard that the ship had started its engines and I asked one of my staff to go over and check to find out what was happening. They were told that the ship was leaving port on an emergency and that we wouldn't be able to have our visit that night so we continued on with our dinner. (When) we got back to our accommodation that night I turned the TV on and we started to watch 9/11 unfolding. I had representatives from America, UK, Israel - so many countries - and we realised straight away that this could have been the start of World War III. We weren't really sure what was happening but we were on the phone straightaway to Defence Headquarters in Canberra. Our aircraft was taken away from us (so) we were on the phone to Qantas immediately, trying get some of the attachés, particularly some of the Americans and Brits, back to Canberra as quickly as possible. We had assistance from the Australian Federal Police

(and) a lot of resources thrown at us if we needed any assistance. The next morning we managed to get several people out. Ansett collapsed twenty-four hours later and I had about two hours' advance notice of that collapse from the Joint Movements Group. I rang Qantas and got the fourteen available seats out of Darwin over the next three days (and) offered those seats to the attachés and their wives. (When) there were no more flights out we tried to see if the Air Force had flights that could pick us up. We were given carte blanche pretty much to hire cars and buses to drive back but I wasn't prepared to take the risk of letting twenty attachés who don't even normally drive (they have drivers of their own) into a hire car and saying, 'Find your way back to Canberra'. I spent quite a lot of time on the phone with two of my staff negotiating with ambassadors and staffs in the Embassies and High Commissions to get people to go round and dig out passports from safes in houses and in the ambassadors' offices. They would airfreight the passports to Darwin Airport and I would go out during the night, pick up the passports, go back to the hotel, pick up the staff, get them to the airport and make sure they got on a plane. We sent them to Singapore or Denpasar or Jakarta even to get them back to Sydney because there's no way we could get a domestic flight back. It took us nine days to reduce the numbers down to zero. I was the last person out. I remember getting on the plane in Darwin absolutely exhausted and getting as far as Brisbane and not feeling 100% and then the Brisbane to Canberra leg was awful because I ended up with food poisoning. I got back to Canberra, lasted about two hours and then ended up in hospital, so I think my resistance level was really down at that stage.
Denise Dunn (enlisted 1976)

I have been on annual camps and exercises to RAAF Base Curtin (WA), Katherine (NT), Townsville and to Cowley Beach near Innisfail (Qld), as well as Nhulunbuy and Kangaroo Flats in NT. In 2000, I was lucky to spend three and a half months with the Australian Rifle Company in Butterworth Malaysia. In 1988, I spent five weeks with the Camel Team as support staff travelling from Charleville in Central Queensland to the Gold Coast. Other highlights include a parachute jump into water, airborne and cliff rappelling. I have flown in RAAF Hercules doing touch and go starts as well as flown over Melbourne standing near the pilot, observing the city of Melbourne below. (I have) travelled by Shark Cat from Sorrento to Williamstown, and seen the city of Melbourne appear

as we travelled closer. (I have) been seasick with the rest of the unit whilst on a patrol boat in Westernport Bay on a weekend exercise and driven from Darwin to Sydney to Melbourne by road, returning from a Kangaroo Exercise. The most frightening experience whilst in the Army was at Puckapunyal when I had to throw a grenade, as it makes a very loud bang when exploding. After all the lessons we all had a turn. I was worried that I wouldn't be able to throw it far enough outside the bunker as I wasn't good at throwing a ball, but the male instructor said it was easy so I believed him, had a few goes and survived the activity.
Liz Stone (enlisted 1967)

One evening when I was on mess duty the cook and I had finished cleaning up after dinner when one of the doctors appeared and demanded a meal. I told him that meals had stopped being served half an hour ago, the food had been thrown out and the cook was just leaving to go home. He told me to tell the cook if the choices on the menu were no longer available he would like a steak. I duly reported this to the cook who had cleaned up, turned off the stove and changed into civvies for the eighteen mile drive home. He lost his temper and said 'I'll give the bastard steak!' He went into the coolroom, cut a slice off a side of beef, re-started the stove, wiped the steak around the slimy inside of the slop bucket then threw it on the stove. He took some greens and mashed potato out of the pig bucket and put them on a plate. When the steak was cooked he said 'Give that to the bastard and I hope it chokes him.' I took it around to the Officers' Mess and put it in front of him. His eyes lit up and he started eating. I had to wait for him to finish his meal and when I went to collect his plate he said, 'Please give my compliments to the cook and tell him I said it is amazing what someone can do when they try'. I made a vow then and there to never get on the wrong side of any cook.
Lyn Kane (enlisted 1964)

While living in at WRAAC School we were sitting in one of the rooms having coffee and one of the corporals went to her room and came racing back and said that there was sawdust on the floor. Somebody moved the mat and we saw that there was a line sawn with some sort of fretsaw and all this sawdust coming up under the mat. The next day she went to the orderly room and reported that she thought there were borers in the door and in the floor. We were having coffee the following night in our pyjamas and we heard a noise which seemed to come from under one of

the huts. We grabbed a torch and shone it under the huts and we saw a ‘peeping Tom’ under there. He scrambled up the other end of the hut followed by us and I yelled out, 'Stop. I’ve got a gun!' Just as he was ready to scale over the fence he just halted. We called the orderly officer and she got him. He turned out to be a groundsman who worked doing lawns and stoking the boilers. So the police came and he was arrested. As I was running the elastic broke in my pyjamas and so I was sort of trying to hang on to my pyjama pants and run and there was much kerfuffle but that was a good time as well.
Julie Elverd (enlisted 1960)

Every two months I would have to go out to Broadmeadows Camp to pick up Modess pads (sanitary napkins) for the girls at 31 WRAAC Barracks. The camp had this great big long shed and I had to go right up to the back corner. On this particular day when I pulled up at the shed, I heard a voice say, 'Can I do something for you, Private?' I said, 'Oh yes. I’ve come for 31’s man hole covers'. I look around at this poor young Lieuy (lieutenant) who had gone as red as a beetroot and he took OFF. Finally I get to the counter and can hear all the boys laughing like hell out the back, having morning tea. I said 'Anybody home or are you all too busy eating and drinking?' They come forward and one bloke says, 'What did you say to him?' I replied, 'I asked for the man hole covers'. He said, 'No wonder he was as red as a beetroot. Now we won’t be able to find him all day. He went out the door and we haven’t seen him since'.
Fran Farrell (enlisted 1963)

Army No. F58054 Rank PTE (L/CPL) Name HEALY W.

AUSTRALIAN MILITARY FORCES

AAB 83b

Revised July, 1952

Record of Service Book

Female Member

SECURITY

All Ranks

"REMEMBER — Never discuss military, naval or air matters in public or with any stranger, no matter to what nationality he or she may belong.

The enemy wants information about you, your unit, your destination. He will do his utmost to discover it.

Keep him in the dark. Gossip on military subjects is highly dangerous to the country, whereas secrecy leads to success.

BE ON YOUR GUARD and report any suspicious individuals."

the WRAAC..

During World War II the Australian Women's Army Service was introduced to release men for overseas service.

The idea worked well, so the Women's Service was brought into being again with the Regular Army in 1951, not only to develop in peace-time a nucleus for expansion in war, but also to release men for overseas duties.

Chapter 7

Shadows of War – Vietnam

> ## *...plays a vital role*
>
> **Today, there are hundreds of women doing important work formerly performed by men and gaining experience which will stand by them all through their lives.**

The Vietnam War cast its long shadow over the whole country. In the community there were strong feelings about Australia's involvement in the war and emotions ran high in demonstrations in the capital cities and letters to the newspapers. The war was brought into people's homes via reports in the daily newspapers and nightly television news coverage. Servicemen and women were also subjected to hostility from civilians. Thousands of people took part in moratorium marches and, as a result, many of the troops departing for overseas were denied the chance to have loved ones farewell them at airports. They were flown out under cover of darkness and returned to Australia under the same conditions.

During those years letters were the lifeline and the only means of communication between the soldiers fighting overseas and loved ones they left behind. For the families at home and the men overseas letters were the link that helped to keep them close to each other. Sometimes it would take several weeks for these precious letters to arrive at their destination. Troops overseas and their families at home were stunned and angered that on one occasion the Postal Union at home, in protest against the war, went on strike and refused to handle mail to and from the troops in Vietnam. The Waterside Workers also refused to load mail, food, equipment, vehicles and other supplies for Vietnam. In some cases, these goods were loaded onto Navy ships by Defence personnel in order to get them dispatched.[18]

Figure 43 Shirley Hall in her room at 31 WRAAC Barracks writing to her fiancé in Vietnam, 1966.

After fighting overseas during their twelve month tour of duty, the troops returned to an Australia deeply divided about the war in Vietnam. Having expected a welcome from the population at home they were often reviled and shunned by civilians who questioned the morality of the war and of Australian troops being sent there to fight. When the soldiers of 1 Battalion RAR (Royal Australian Regiment) marched through the streets of Sydney on their return home from Vietnam in 1966 a female protestor threw red paint over the commanding officer and others leading the parade. I saw this incident on the evening news and was appalled that our soldiers had been treated in this manner. I have sometimes thought about that woman and wondered whether, with

maturity, she had reflected on her actions and felt embarrassed or ashamed of what she did. After all, the decision to send troops to Vietnam had been made by the governments and not by the soldiers themselves.
Lyn Kane Dale

On our return from Corporals' (Promotion) course in Sydney, Peta Strauch and I had to wear our uniforms although we had informed the powers that be there was a moratorium march scheduled for that day and it would be risky being in uniform. We were picked up by the duty wheels at the airport and driven straight to the front gates of Watsonia Barracks even though we had told the driver – male – to take us in through the back. Imagine our horror to find a large crowd of people shouting and waving placards in front of the locked gates. Before the driver could reverse we were attacked by people pushing and rocking the vehicle and thumping on the bodywork. Eventually we made it out of there and around to a side gate and entered the barracks shaken but unhurt.
Hilary Jones (enlisted 1970)

In 1966 I was working at 3 Base Ordnance Depot at South Kensington in Melbourne and we were very busy supplying units which were going on military exercises before leaving for the Vietnam War and others who were preparing for their twelve months' deployment overseas. On several occasions while driving through the city in uniform young people, mostly protesting university students, would shout at us 'Get out of Vietnam. Stop the war'. And sometimes we would shout back 'Get a haircut and a job!' Hundreds and sometimes thousands of people took part in moratorium marches in the cities. One day after work another WRAAC and I decided to stop off in the city to do some shopping instead of going straight back to the barracks. We walked along Bourke Street to Myers to find hundreds of protestors, many of them university students, making their way along the footpath. The line parted to allow shoppers access to the stores and as my companion and I went to enter the shop they let her pass but one of the young protestors grabbed my arm and said 'Come over here. We need to point out to you a few facts about the

war'. I said 'Please let go of my arm' and he replied 'I am not going to let go of your arm until you listen to me. You people are in the wrong for murdering innocent women and children in Vietnam'. People in the crowd were chanting anti-war slogans and several other young men around him also angrily yelled at me about American atrocities towards Vietnamese civilians. Again I asked him to let go of my arm and he again refused, saying that Australians should not be fighting and killing in Vietnam. He then said the protestors were going to take me along with them. I was quite frightened as he was much taller and bigger than me but I was determined that he would not see how scared I was. All of this took place in just a couple of minutes but it seemed to last forever. Finally I said 'If you do anything to me you will get into a lot of trouble. Now, let me go.' One of the other protestors stepped in and said 'She's right; you will get into strife over this. Let her go, mate,' and he did. I walked into the store to where my companion was standing and when I got around a corner where I could no longer be seen by the protestors I started to shake uncontrollably. My legs would not hold me up and I collapsed into a chair where I sat for about five minutes shaking like a leaf before I was able to stand again. After several incidents like that in Melbourne, WRAAC personnel were warned not to stop off in the city after work while in uniform. Although we still travelled by public transport we were told to get off the train and straight onto a tram back home to the barracks.
Lyn Kane (enlisted 1964)

We had two children. I stayed in Holsworthy married quarters when John went to South Vietnam in 1967. (Because) of very nasty protestors, security was so tight at Mascot Airport that the men were secreted out of the country at dawn and then on return the same applied! It was so tragic to see these young men treated so badly for nearly thirty years before the Government recognised their plight. It was a very hard life at times with John and his undiagnosed Post Traumatic Stress Disorder (PTSD) but love prevailed and we remained together until his death from cancer in 2003. Our son, Stephen, joined the Army as a Sig in 1978. Two of my grandsons joined the Navy and one is still serving at HMAS Stirling, WA after two trips overseas.
Wilma Stralow (enlisted 1955)

I was sent to Puckapunyal for a clerical course, then back to Victoria Barracks and was put into the National Service intake section. My duties were collating all the paperwork for all the inductees that were coming in during the sessions. I was fortunate enough to be at the Centre when singer, Normie Rowe, came in so there were a lot of publicity people there and a lot of police and army personnel because there was so much kerfuffle (anti-war protests) going on.
Desiree Harms (enlisted 1965)

In 2013 I was surprised to find out about controversy surrounding the purportedly 'fixed' conscription of singer Normie Rowe. According to published reports, the son of my old 1964 Army Public Relations boss in Melbourne, Lt Colonel Beverly Coultman-Smith, claimed that his father had admitted on his deathbed in 1978 that it had been his idea to draft Rowe. He had been hoping for a public relations coup like the unprecedented publicity that followed Elvis Presley's enlistment into the US Army. 'Dad asked me if I ever saw Normie, to apologise,' Victoria Police sergeant Grant Coultman-Smith said.[19]
Helen McLeod (enlisted 1963)

Most of my time in WA was working in Hay Street in the city at the Combined Services Recruiting Centre and I had to walk there from Swan Barracks each day. There was so much negativity towards the Army because of conscription and Vietnam and I felt very conscious of walking through the city in uniform. I always wore my sunglasses to avoid eye contact with people looking at me. Because of the Army stigma I also avoided telling any civilians I was in the Army. I used to say that I was a Commonwealth Public Servant and if really pushed would say I worked for Defence. Socially it was so much easier to stick closely with my fellow Army members. It is also such a way of life. I had been selected to go to WRAAC Officer Cadet School (OCS) for Officer Training but chose marriage (in 1974) and a posting to Army Welfare South Australia instead.
Jayne Rickard (enlisted 1969)

During a pre-embarkation party for John's unit leaving for Vietnam, one chap decided to be a bit too crude and started to talk in Morse code – thinking none of the wives and girlfriends would understand. As I had not long before this done my Operator Keyboard Radio (OKR) course

and knew Morse code, John asked him to stop and then spoke to the host when this chap continued, and let the host know that I understood. I felt bad that the chap was asked to leave but the host was most apologetic as he did not realise that I had done the course. The camaraderie of John's unit has been wonderful. We would meet up each year somewhere to enjoy Anzac Day and reunions. Since John passed I have continued to go to reunions and am always well looked after.
Daphne Gardner (enlisted 1964)

One of my positions within 2 BOD (NSW) was working in Camp Earmark which held and controlled the loan stores for units going on annual camps and the major exercises the Army held over the year, particularly getting troops up to strength ready for Vietnam, so I got to meet lots of different people (mainly men). One of the 'highlights' was when 1RAR (Royal Australian Regiment) colours (regimental flag) were paraded into the depot to be 'laid up' for storage before they went to Vietnam. The troops' personal effects were also stored in the depot. When there were casualties their personal effects had to be checked before they were forwarded on to the Next of Kin (NOK).
Merle Viles (enlisted 1962)

The 7th Battalion Royal Australian Regiment was to march through the City of Melbourne before their twelve month deployment to Vietnam and Noel asked me if I could get off work to come and see the parade as he was leading his company and saluting the governor. I managed to get an early leave pass, met my room mate, Shirley Hall, in the city and we found a vantage point near the dais. As the battalion marched past and Noel called 'eyes right', all heads including his turned in our direction and he looked straight at me as he saluted. It was one of those moments frozen in time -very exciting and romantic. After the parade dispersed we met and talked for a few minutes. That was the last time I saw him before the battalion went to Vietnam. We had discussed marriage but I had doubts that I would make a very good officer's wife and he planned to be a career soldier. My thought was that when I got out of the Army I wanted to be out and not have my future controlled by the Army and my husband's commanding officer.
Lyn Kane (enlisted 1964)

Sydney Airport: a hot summer day in 1968

I see myself, just turned twenty-one, standing with my parents and some friends, tears not far away, waving goodbye to my brand new husband. I see Jim, just twenty-three, clad in jungle greens, clutching a kitbag, simultaneously anxious to leave and wanting to stay. I gave him a kiss and a hug, and then watched him walk out to the plane, wondering, in a small, scared, private place if I would ever see him again.

I actually found myself crying as I wrote this. Of course I do cry easily (for both good and sad things) but that memory came flooding back as if it had occurred a week earlier. I recall turning around to my Dad, who was always a source of support and comfort, and being stunned that he too was crying. He said 'I thought that we went to fight a war so that this NEVER happened again, and now another generation is being sent to another war we still don't understand.' For months we relied on those letters c/- of an Army Post Office (APO) to convey our thoughts, hopes and prayers to those young men living in a land we saw depicted nightly on television: a totally foreign land, for all that it dominated our lives through the media. In due course, most of those young men came home to us, some more damaged physically and emotionally than others, but none untouched by the experience. For those who returned in the black anonymity of a body bag and Army issue coffin we reserve the strongest emotions of all. We've gone on to raise children, we've grown older and greyer, had our fair share of smiles and tears, and yet they remain as once we knew them, magically preserved in a time when we too were young, with everything ahead of us. A time when everything was possible and middle age was a term that applied to our parents.
Kim Lake (enlisted 1966)

1 Australian Field Hospital in Vietnam had the highest standard I have ever encountered. It was clean, air-conditioned and purpose-built with unlimited staff who were highly motivated, capable, willing and subject to military law which is a great motivating force. The barracks were spartan - hot and full of sand. We had an air-conditioned night duty room which slept two nurses. We were small by comparison with the Americans and we received invaluable support from the Americans. Wounds were caused by gunshot, mines, grenades, beer cans, motor vehicles, ditched helicopters, kitchen knives, you name it. Other wounds were caused when some of the soldiers gave a small brown bear, called a

Honey Bear, some beer. It got drunk and bit several of them, who then had to have rabies shots as all animals in South East Asia were suspected of carrying rabies.
Ann Hall (enlisted 1967)

We were married only a couple of months and Greg went to Vietnam. After I got out of the Army I moved into a hostel close to where I worked as a dental nurse. Because of National Servicemen being sent to Vietnam there was a lot of hostility around the society at the time. When some people found out Greg had been an enlisted Army apprentice they said that he deserved to die. When Greg was due to come home I arranged to meet him in Sydney and then we would go up to Brisbane. I said goodbye to everybody in Melbourne and got on the train. During the night the train came to a standstill and the porter came and knocked on my door and said 'Your husband's not going to be coming home.' I didn't really know what had happened to him. Your mind goes through all different things. What's happened to him? Was he killed? It was such a shock because I was on my way to meet him and had heard nothing.[20]

I decided I would go on to Brisbane so when I got to Sydney I telephoned my father and step-mother and asked if I could go there for a few weeks until Greg finally came home from Vietnam, but I had no contact with Army. The first contact I'd had from anybody was that night on the train to Sydney.
Ruth Grace (enlisted 1964)

Arthur left Australia on 26 April 1967 and our daughter Tracey was born on the 27th, so he didn't actually get to see her in person but lots of photos were sent to Vietnam. I couldn't understand why the Army didn't keep him back till she was born but it was only later I found that he didn't apply to stay back, the reason being that he was a welfare child and his biggest aim in life was to have his own family. He told people if he'd seen his daughter or son he would never have gone. Now when you're a regular soldier you don't have a choice. You've got to front up and do your bit so he figured it would be better if he went without seeing her. And then we got the word that he was a very sick boy with malaria in December of 1967 and so we went through five days of receiving telegrams saying 'Seriously ill'. Later in the day you'd get 'Very seriously ill', the next day back to 'Seriously ill'. Then we had the dreaded knock at the door and the chaplain came to tell us that Arthur had died on 5 December.

I was a person that could be psychic at times. I'm not really sure on the time but it probably would have been about two months before Arthur died that I was asleep one moment and the next I was yelling. My aunty came racing out and was shaking me and asking, 'What's going on? What's going on?' I said 'He's not coming home.' She said 'What do you mean?' I said 'Arthur's not coming home.' She said 'How can you know that?' I said 'I don't know but I just know it.' Oh, I can't explain how that felt. I think I cried for about four days. I must admit any time I get a psychic feeling now I actually block it. I could be getting happy psychic feelings but I'm not game to take the risk.
Jenny McLay (enlisted 1962)

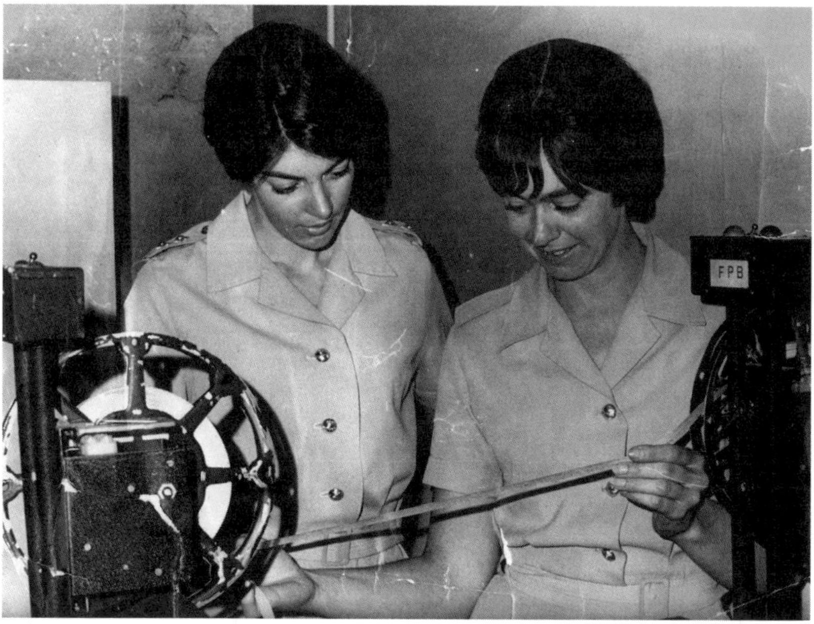

Figure 44 Julie Elverd (right)

We used to receive the signals tapes manually and some of them were just cut out holes on tapes with no writing but a lot of them had the actual printing underneath the tapes. We used to stand there and check the tapes for legibility and clarity and quite often we'd see some of our comrades who had left the units and gone to Vietnam. With the fatal

casualties, quite a few times we saw our mates and we couldn't say anything but it was hard to see 'so and so' had been killed or 'so and so' had been wounded. We actually put our hands up and said 'We want to go to Vietnam with our colleagues', and no, it wasn't allowed and there was no way we were going to go there.
Julie Elverd (enlisted 1960)

My last posting in April 1969 was to HQ Tasmania Command in Hobart as Orderly Room Ration Clerk then Pay Rep. I was Pay Rep towards the end of National Service and the return home of our troops from Vietnam. Part of my duties involved going to the Repatriation Hospital to pay our soldiers. At times this was a very sad occasion to see firsthand the injuries our soldiers had sustained. Also, being a Tasmanian, I encountered a lot of guys I had grown up with.
Andrea Woods (enlisted 1967)

Colin served in Vietnam for twelve months, from January 1970 to January 1971. At the time of his posting, we had a two-and-a-half year old daughter and were living in Melbourne, away from my family. It was a long and lonely twelve months. Being in Signal Corps, Colin's time in Vietnam was probably a lot less stressful than that of an Infantryman, for example, so he came home to me in relatively the same condition as he left, and suffered no ongoing ill effects as a result.
Marie Taylor (enlisted 1962)

Parties at Vung Tau, Vietnam were lots of fun. There would be twelve women and eight thousand men. The odds would never be that good again. We were going to have a good time (but) I felt I was well-liked and respected. We went swimming, sailing, shopping at the PX (American Army store) and in town. Chanel perfume was duty free, Oroton bags - I had six! We would chopper up to Nui Dat for lunch but not often, for we nurses worked six days a week.
Ann Hall (enlisted 1967)

We married twelve weeks before he went to Vietnam and (I) wrote him every second day. The boy who left was not the man who came home and the girl he left behind was not the woman he met! (It was) to hell and back for some. My Dad was a psychiatrist, a Military Cross winner from the Burma campaign (World War II) and a consultant in the Repat

(Repatriation Department) so I guess I was better prepared for what was to come in some ways and not at all in others. The question they should ask a young man is not 'Are you prepared to die for your country?' but rather 'Are you prepared to kill?' That is what destroys their soul!
Joie Norton-Willson Scheinpflug (enlisted 1967)

John and I did not have a long time between my course finishing and us getting married but we still managed communications by letters. We continued letter writing while John was in Vietnam from May 1966 to May 1967. We numbered our letters and at times they would come in bulk so the numbering system allowed us to read things in order. But it was very frustrating when no letters came.[15]

There was no assistance from any welfare group until I had to inform John that we had lost the baby we were expecting. It caused John a great deal of heartache and he was ready to go Absent Without Leave (AWOL) to try to get back to me. Fortunately some of his mates knew about the baby and got him through that stage. I was finally contacted by Welfare and they got some sort of message to John that I was recovering. John came back a different man from the happy-go-lucky chap he was when he left. Being a young bride of just over two months when my husband left for Vietnam and only twenty years of age when he returned, it was hard to know what to do for the person John now was. I feel that my training as a nurse as well as the training in the WRAAC helped me to manage.

We went on to have two sons and a daughter but as they grew up John was a very strict father towards them. Now they look back on those days and say that Dad's discipline has helped them in their lives, although at the time they were typical children and did not like the restriction. Our sons went on to join the Army and RAAF so it did not scare the boys away from that life and one grandson is now waiting for enlistment in the RAAF. It was very hard at times to stay but I feel a change in John came one night when he was going to drive away while drunk and arguing. I stood at the back of the car and told him that if he left, when he returned, if he returned by not having an accident, the boys and I would not be there. John did not leave! We went on to have our daughter, seven grandchildren and now one great grandson. Unfortunately John did not see our great grandson as he passed away in 2010 from

complications from Parkinson's disease. We had been married for forty-four and a half years. I miss him each and every day.
Daphne Gardner (enlisted 1964)

I was discharged in February 1966 with the intention of getting married. I met his mother over the Easter period and was to return home to NSW, but we got a telegram (no emails or SMS in those days) saying that he was posted to Vietnam. I quickly arranged for his mother and me to get to Sydney to see him before he left. I had to go out and buy my own engagement ring and we got engaged the night before he sailed. He left on the HMAS Sydney (nicknamed the 'Vung Tau Ferry') taking over all sorts of trucks and vehicles. It was the first shipload of vehicles to be sent over and the deck was covered with them. My parents, his mother and I all went down to see them off. I think it was Garden Island that they sailed from in those days. We all had streamers and there were many family and friends there for all the diggers and sailors. As the ship slowly pulled out the streamers started breaking. My dad went along picking up the broken streamers and we were tying them together. Ours was the last streamer to break. John was in Vietnam for fourteen months with HQ 1 Task Force. John was there for all the aerial spraying that went on and was carrying ammo to the helicopters for the Battle of Long Tan in which eighteen Australian soldiers were killed and twenty-four wounded. His protection in those days was a sandbag on the floor of the jeep/truck he was driving! Luck was on his side as he straddled an unexploded mortar on one of his many trips. As he and his mate were last minute replacements, they were forgotten when the rest of the boys were sent home after twelve months. They finally made it home in May 1967, two days before his mate's wedding, but only just, as the plane sent to fly them home burst a tyre upon landing and they had to wait for it to be replaced. When they reached Singapore they were told to change into civvies as they didn't want them landing in Melbourne in uniform. Still he came home safe and we were married for forty-four years before he died suddenly, from an aggressive throat and lung cancer, in March 2011. Another one for the Cancer Register.
Jean Fullerton (enlisted 1963)

We have certainly had our ups and downs. I don't think anyone returns from a war unaffected, but we've hung on. He is on a vet's disability pension but is better off than a lot of other vets.
Elizabeth Anderson (enlisted 1964)

Mike went to Vietnam a year after we were married and I returned to my parents. I wrote to him every day. I even numbered the letters as the mail often got messed up. At one stage the posties in Sydney went on strike and wouldn't load the mail to Vietnam. The fellows had a poster made telling everyone to 'Punch a Postie' when they got home. Mike wrote as often as he could. I still have about three hundred of his letters in a box. What a terrible year without Mike. My father died suddenly, aged fifty-four. I thought having Mike home would help but he came back with his own problems and so started many years of good times and also trials. We have been on our own for the past fifteen years, we have built our own home, which is still standing, and have become even closer than before. I am proud of Mike for accepting help after being diagnosed with PTSD and depression; it has made all the difference. I thank God for helping us through and blessing us with all these years together, and also for our two children and six grandchildren.
Lorraine Field (enlisted 1964)

I was engaged several times while serving but they fizzled out. When I took my discharge I moved to Wodonga, Victoria, and worked as a civvy doing the same work as I had been doing in 2 BOD (2 Base Ordnance Depot) – re-supply for South Vietnam. I met my husband soon after I moved. He was a RAEME (Royal Australian Electrical and Mechanical Engineer) sergeant just returned from South Vietnam. I was twenty-seven and he thirty-one. He had not been married before. He was promoted to a Warrant Officer Class II (WOII) not long after we married, and, as I was used to obeying Army orders, I was easily bullied and life was very difficult. We had two young daughters, both with medical problems from Agent Orange. After his discharge when he finished his twenty-year service in 1975 he struggled with civvy life, eventually being diagnosed with PTSD in 2000. After several suicide attempts with all the other stuff that goes with it, he passed away in 2008 of heart failure. We were married for thirty-eight years but the last few years were difficult for all of us. I did receive much help from the Vietnam Counselling Service, during the years from the late 1990s onwards. I have no desire to

remarry and try to be a 'modest' widow and help my daughters and their husbands. They are not going to have children.
Merle Viles (enlisted 1962)

I felt I was unfortunate to be a soldier at a time when women were required to leave the Services if they wanted to marry, so it was an agonising decision to request a discharge to marry my fiancé a few weeks before he was posted to Vietnam in 1966. Like many others in my position, I found that I hardly knew the stranger who returned to me from Vietnam. We settled in South Australia to be near his family, where we battled to live with the terrible anger that he just couldn't keep under control, his denial that anything was wrong with him and the difficult lives we led because he was unable to fit easily back into society.
Helen McLeod (enlisted 1963)

I married a soldier who was posted to Vietnam for a one year tour of duty, leaving me with small children to take care of in Army married quarters away from family and friends. While he was overseas I felt very isolated and received no support from other Army wives who gave me a hard time about sending my children to a Catholic school. After my husband's return from his deployment in Vietnam there were problems with his drinking and depression. After twenty-one years, four children and having suffered thirteen miscarriages I divorced him. Through my job as a real estate agent I met my second husband and we have been happily married for over twenty years.
Trish Courtney (enlisted 1962)

I did my basic nursing training at the School of Army Health, Healesville, Victoria, where I met my future husband, who later went to Vietnam. After the training course finished I was posted to Puckapunyal, Vic. During my time there my future husband drove all night to come and see me to put an engagement ring on my finger. About two months later at Easter we went on leave where he put me in the family way, which resulted in ending my Army career.[21] After I found out that I was pregnant, Bob wanted to marry me as soon as possible, but his parents would not give permission, their reason being 'You are too young and she is a Pat' (Catholic), so Bob, being an adopted child, went to see the Chief Secretary of the South Australian cabinet, Sir Lyell McEwin. He also got permission from the women police and, of course, his CO and, after intense questioning, he got his permission. Bob had already made arrangements with an Army padre for a wedding date; so on the Saturday before 1st of July 1961, he rang me at my mother's place and said, 'Get your wedding dress ready. We are getting married next Saturday'. Meanwhile, my mother had a plan for me as well. She had already signed me into Fullarton Catholic Home for Unmarried Mothers (Adelaide), and I was about to go in there when Bob rang me. Lucky for me. I found out later that many who went into homes for unmarried mothers were coerced or forced to give up their babies for adoption. I remember telling them I wanted to keep my baby. I am sure all young girls would have said the same.

I had my fourth son on 16 January 1969, the day before Bob set sail on HMAS Sydney headed for Vietnam. We communicated via telegram. He saw his baby son, Phillip, seven months later when he came home for five days Rest and Recreation (R&R). I was living in Salisbury North, SA, in a housing trust area amongst hostile people who didn't understand why our men went to war. I was alone bringing up my boys and I could not cope and as a result suffered a nervous breakdown. The Family Liaison Officer had Bob brought home which cut his tour of duty short by a couple of months. I was on anti-depressants for ten years but I am now fully recovered. Bob was affected badly by his overseas service and changed so much after Vietnam that we separated in August 1973 and divorced three years later.

In 1977 I married again and had my beautiful daughter, Heather. However, my second husband turned out to be violent towards me and my boys so we divorced in 1983. Bob had also remarried but we always kept in contact with each other in order for him to see our four sons. Bob's second wife was a violent woman who was extremely jealous of me because Bob continuously talked about me (he never stopped loving me). After my divorce he divorced his second wife and then he rang every night and visited me on weekends. I would say, 'Don't put your hopes on me', but, finally I agreed to reconcile with him. We courted secretly for months. I wanted it that way just in case I couldn't go through with it. I didn't want to hurt or humiliate him. In August 1983 Bob, my eight year old daughter and myself moved to Semaphore, SA, exactly ten years almost to the day of our separation. It wasn't planned, it just happened - fate I suppose. Bob treated Heather as his own child and to

this day says she is his daughter and includes her in everything along with our four wonderful sons. We remarried in Adelaide in 1984 and now have nineteen grandchildren and nine great grandchildren. Bob and I are still married but after he went to Vietnam his personality changed dramatically. He suffers from PTSD, also anxiety, depression and alcohol abuse. He is now Totally & Permanently Incapacitated (TPI) and under constant psychiatric treatment. A lot of things happened to both of us during our ten year separation, but I found the grass was not greener on the other side of the fence and that I am better off with the devil I know than the devil I don't know.
Gloria Coombe (enlisted 1960)

I served only twenty months of a three-year enlistment due to the legislation of the time that if you were getting married you had to be discharged. It was so unfair. Speaking for myself, I was not ready to be discharged, but my fiancé was going to Vietnam. I was discharged on 26 September 1969, my twentieth birthday, but it turned out that he didn't go to Vietnam until October of 1970. I think that all the money, time and effort put into training us was wasted, especially knowing husbands would be away for twelve months, which would have given most of us the opportunity to finish our enlistment. That marriage ended in divorce after twelve years for my safety and sanity. We had four children and all of them witnessed domestic violence. We were living in married quarters at Puckapunyal and once, after he was violent towards me and I threatened to leave, he spoke to a counsellor and the Army sent someone around to talk to me and tell me I hadn't been a good wife. The Army attitude of the day towards the 'housewife' a soldier married seemed to me to be: 'If the Army had wanted you to have a wife you would have been issued with one'.[22]

Six years later I married again, to another Vietnam veteran, and that relationship lasted two months shy of thirty years at his death at the age of sixty-nine in November 2013. He had served overseas in the Malayan Emergency, Indonesia Confrontation and two tours of Vietnam. I was still at school when he went to Vietnam the first time. He was wounded in Vietnam and medevaced[23] to an American hospital where the highlight for him was being presented with a Purple Heart by Nancy Sinatra. It was quickly taken back after she left the ward because he was an Aussie. We had one child together. She was born deaf in one ear. Yes,

he did have a lot of medical issues. It was not an easy road living with a Vietnam veteran with PTSD, which is the current name for what was known in World War I as shell shock and in World War II as battle fatigue. I think all wives travelled a similar road from all wars Australian Servicemen participated in. Vietnam was different. The boys didn't come back to a grateful nation that had sacrificed its youth but to an Australia that was embarrassed and ashamed of our young men, which exacerbated the trauma experienced by many. PTSD doesn't just affect the soldier. My experience tells me that the wife and kids suffer too. There are generational problems and issues, especially in the physical and mental capacity of (some of) their offspring. Socially, many veterans are withdrawn, alcohol becomes a big problem, mental issues arise, and there are domestic violence issues and other underlying problems that (can) lead to the irretrievable breakdown of marriages. Families decline and disassociate and the ugly curse of PTSD continues.
Trish Webb (enlisted 1968)

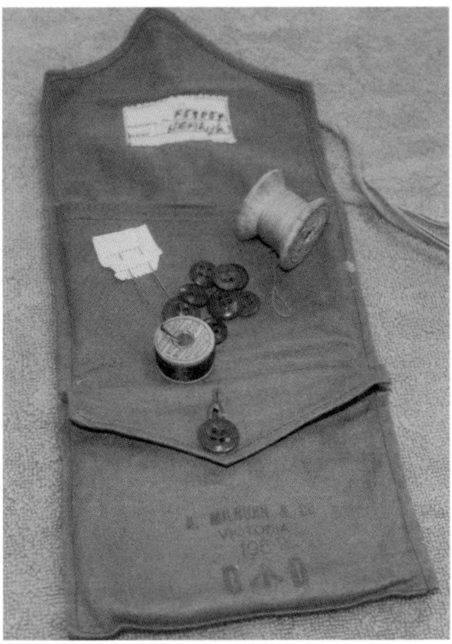

Figure 45 Housewife kit

There was a group of nine women I didn't know closely but their husbands had been away (to Vietnam). There was a huge amount of miscarriages between them. Every one of them had had a miscarriage. At that time I had my firstborn, Benjamin, but between Benjamin and the next child I had four miscarriages. In all I had thirteen pregnancies but there were only four successful ones, so I had my four children. I doubt if there has been any study made but it just seemed to me that the couples who had not had children before the husband went to Vietnam had so many problems when he returned conceiving or carrying to term. Those of us who had a child quite easily before our husbands went did have problems after they came back. So it didn't take much for us to come to the conclusion that Agent Orange and all the other defoliants and pesticides that were used there had an effect on the husband and therefore on the wife. It was a very tragic and sad thing to go through and there were so many of us, which hasn't actually been acknowledged by anybody, even to this date.

My first marriage did not survive. When my husband came back from Vietnam he was very different. He seemed not to want to be married, even though we had three children. He didn't take any notice of them. He was unfaithful, verbally abusive, emotionally abusive, and on occasions, physically abusive to me. I put up with it for quite some time, but after eleven years it was too much and we parted. To me it was a failure. I'd always been brought up – you make your bed, you lie on it. However, a man who ignores his wife and children, plays around and spends money on other women is not a good role model for three boys. Ultimately I believed the best thing was to have one almost rational parent looking after them and teaching them values than for them to be in an abusive, or witnessing an abusive, relationship which would leave them scarred. We came back to Mt Gambier to live, where the children attended school. After some time I met my present husband and we married and had a daughter. He is a good man and he brought up my sons as his own. And with our boys and our daughter we have a good family and two lovely grandchildren.
Sheila Cusack (enlisted 1964)

A signal to Vietnam— it's a boy!

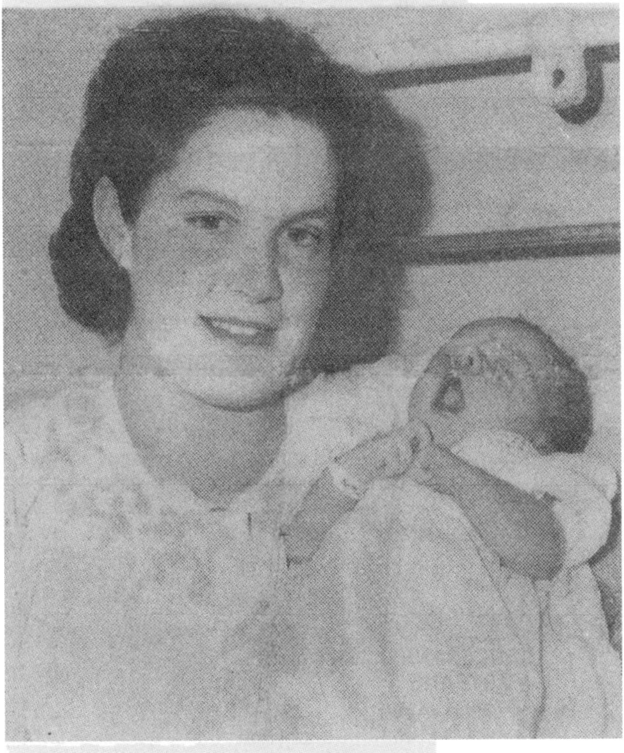

DON'T YAWN NOW, son, this picture's for your Dad! — Here's a signal for Signaller Ronald Minton, serving with the Australian Forces in Vietnam. Your wife, Sheila, and your new son are doing well.

The baby was born two weeks ago at Airlie Hospital, Ivanhoe, and weighed 7 lb, 4 oz.

Signaller Minton (below) has been overseas for seven months. First glimpse he will have of his son will be his picture.

We had been married seven years and Ken was diagnosed with melanoma. We had been posted to Townsville and it was two years into the posting. He had a very rough time of it as it started behind his ear and went down to his glands. After lots of operations and treatment he was retrained in administration and posted to Adelaide. This was a very low period of our marriage as he went from being a soldier to an office worker virtually overnight. He never felt accepted after that. You cannot take the soldier out of the man. He has suffered PTSD (Post Traumatic Stress Disorder) for over thirty years and it was only diagnosed five to six years ago. It became much worse when he left the Army. He is now on medication that has made a huge difference, although we still have some very down times. We know what to expect and adjust accordingly.
Fran Sinclair (enlisted 1972)

My dad was a Vietnam vet and my parents' marriage broke down when Dad was about forty years old. I was in the Army (then) and I was nineteen years old. My marriage broke down after thirteen years but I knew I shouldn't have married him in the first place.
Sandy Jarrett (enlisted 1980)

I was married to Bob for thirty-two years and he was in the Army for most of those years. When Bob finally discharged from the Army he decided to join the Reserves. He was talking to a female warrant officer and said 'Betty would like to join but she's over the entry age'. This warrant officer said, 'She has had previous service so she can join'. I was over the moon. This was the Army I wanted to join in the sixties. Drive trucks, fire weapons, go bush. Oh, I was rapt.

After her experience as an Army driver Betty continued driving trucks for the rest of her working life, often long distance and interstate driving.

Like it was for a lot of drivers this played havoc with married life and I did love driving so much I wasn't prepared to miss out. Consequently I divorced in 1998 and was able to drive the hours without worrying about going home and putting up with the mood swings of a Vietnam vet. Depression is a problem with a lot of these guys and they dislike taking medication for it.
Betty Hall (enlisted 1963)

Marion Quick had a brief romance with a recruit while she was a nurse at Puckapunyal.

We kept in touch over the years. He came back from Vietnam, got married and had a couple of children. That marriage failed as did many marriages through that time because when the guys came back from Vietnam they had no support, no counselling and it placed a lot of stress on their marriages.
Marion Quick (enlisted 1965)

After my marriage I took a casual job with an organisation which assisted veterans in obtaining the compensation that they needed to survive. My role was to assist in finding employment for them and their children. It was a rewarding time for me, because I felt I could assist the veterans from Vietnam, Korea and earlier conflicts. I also felt that talking to them and being friendly was good for them, to relax them before they were interviewed by my manager. Many have PTSD, and some days I knew that they were having issues. Many would just get their gear and go into the bush to re-adjust their lives accordingly. I have many friends whose husbands are Vietnam veterans and I have been amazed at how many of the wives have travelled through their problems. It would have to be a very special woman to understand. I have seen some veterans destroy their wife's life, divorce her, and take up with another woman, younger in some cases, and as the man's health deteriorates the new wife ends up with all the Department of Veterans' Affairs (DVA) benefits. This saddens me a lot.
Denise Darbey (enlisted 1973)

I met my husband, a Vietnam veteran, in Perth in 1968. We married in 1969, had two daughters and spent most of our married life in mining towns in Western Australia. There were some happy years but eventually his depression and heavy drinking took its toll and after twelve years the marriage crumbled. One day one of the children found him sitting on the floor in the bedroom holding a gun to his head. He refused to get help or counselling. I took my daughters and moved to Perth where we stayed for three months. He gave up drinking and promised all would be wonderful if I just came back so I felt I owed it to all of us to give it another try. The kids and I went back, he started drinking again and another day, after I disagreed with him about something, he got explosives (he was in the mining game so had access to them) and threatened to

blow up the house and himself in it. I could see there was no future for us and I was worried about the children's safety. He would not have deliberately harmed us but I couldn't take the chance. Once more I packed up the kids and moved to Perth – this time permanently.
Lyn Kane (enlisted 1964)

I had a long distance relationship and to get together we 'had to get married'. I told the girls in the mess if it did not work out I would get divorced! What a simplistic view. What a mistake really. I had not had more than a total of a fortnight with the man when we got married. I was the young, trophy wife and he was a divorced Vietnam vet Infantry Major and it did not take me long to work out why his first wife left. He was volatile, angry, heavy drinking and unkind. He was violent within six months when I took a wrong turn driving the car home after a mess function. I did not have the courage to leave at that point. I was so worried about what people would think. I went on to have three children and spent a lot of time thinking about leaving. I was unable to raise my children the way I wanted to in that marriage. I developed a plan of upgrading my social work qualification to a degree so I could earn the most money possible to support my family. I established myself professionally and finally after sixteen years I initiated divorce. In the process of separation he threatened to shoot the children and me. After our divorce he was diagnosed with PTSD. He has a poor relationship with our children and was never very involved in their lives. After divorcing in 1992 I met my wonderful husband, Locky, in late 2004. His wife of thirty years had died in early 2003. He'd had a very happy marriage and that is what he has offered me. He is a very kind, loving and gentle soul and our adult blended families all get along very well. I do feel very blessed.
Jayne Rickard (enlisted 1969)

Chapter 8

Marriage and afterwards

From the 1950s to the late 1960s women could not remain in the Army after they married. A great deal of money, time and energy was spent on training, clothing, housing and feeding them, and by allowing them to continue to serve after marriage some of that expense would have been recouped. The majority of women I spoke to who served in the 1950s and 1960s expressed sadness that when they chose to marry they were no longer able to stay in the Army.

A large number of female soldiers, because of their close proximity to male soldiers both during working hours and also in leisure time, married Vietnam veterans and this book seems to contain a very high proportion of their stories. I did not pick and choose women to interview. These women chose to share their stories with me. When they heard I was looking for ex-WRAAC and ex-RAANC members to talk of their Army experiences the stories came flooding in. It is significant that the women interviewed had strikingly similar stories to tell and were eager to share them with others.

These are heartfelt stories of the courage and resilience of these women, most of whom were teenagers when they enlisted and were often very young when they married, in some cases just weeks before their husbands went off to war. They had to deal with life on the home front, loneliness and worrying about their husbands who were away fighting. A large number of young wives had to deal with pregnancies, miscarriages, childbirth, and taking care of young children on their own. They were often living in Army married quarters many miles away from the support of close family and friends, and they missed the friendship and camaraderie of their Army friends.

After their men returned from overseas, many young wives had to adjust to living with a very different man to the carefree young husband they had married. Some wives struggled to bring up families under tough conditions and found themselves supporting husbands suffering from Post-Traumatic Stress Disorder (PTSD), depression, nightmares, anxiety, alcohol abuse and, in some cases, domestic abuse - the effects of exposure to Agent Orange, the defoliant sprayed on the jungles of South Vietnam during the war.[24] Many women dealt with difficult and frightening experiences on their own and tried to protect their children from witnessing (and in some cases experiencing) domestic violence and intimidation in their homes. Some wives, realising that their husbands had no intention of seeking professional help, felt that they had no other option than to leave the marriage for their own and for the sake of their children. After spending years in denial, many of the men finally acknowledged that they needed help and sought medical advice.

In August 2013, during a reunion of former female soldiers at the former 31 WRAAC Barracks[25] in St Kilda Road, Melbourne, the *Lady Soldiers* film was screened. A number of the forty ex-WRAAC and ex-RAANC attending became quite emotional and offered their stories for addition to this book. This opportunity has given them a chance to tell their stories of an era in Australia's not too distant past. At the dinner I sat at a table for ten and discovered that five out of the ten women at my table had been married to Vietnam veterans who had all suffered PTSD and other problems from their service in Vietnam. They told me it had been very cathartic for them to be able to see the film and to speak about their own lives to others who could relate to them and who had lived through many of the same experiences.
Lyn Kane Dale

We had a lot of recruits coming through 3 Camp Hospital, Victoria. It was forbidden for us to fraternise. However I had a brief romance with one of them. We obviously had to meet in secret – no-one knew about it. It lasted several weeks and then he had his march out and left Puckapunyal and went to Vietnam.[26] He wrote a couple of letters but by then I was dating the man that was to eventually become my husband.
Marion Quick (enlisted 1965)

We used to go over to Balcombe to dance lessons as the apprentices needed partners to learn to dance so it was good for us too. I went to one of the dances and this guy caught my eye. He had gorgeous eyes. When I danced with him he said, 'I bet I can tell you what perfume you're wearing'. Mmm I thought – Smart Alec Bob. He did pick the name of it but then informed me that his mother used to wear it too. I didn't know whether to believe him or not but I was very smitten. I found out later that yes, she did wear it. I ended up marrying Bob. In fact he was younger than I and when we did end up becoming more involved and I was pregnant because he was under legal age we had to get the court's permission to marry. We were really sure that is what we wanted. The padre went in to bat for us, saying he thought we were mature enough to cope. Of course this meant I had to be discharged from the WRAAC which was a disappointment, but these were the rules and we couldn't change them.
Betty Hall (enlisted 1963)

I had a few brief romances during my Army life. This all changed on my posting to Tasmania Command where I met my husband. Wayne was an ex-Army apprentice and also a Tasmanian. Another WRAAC and I were lucky enough to find two units together and moved out of barracks. Both of us had an upcoming marriage so our partners were to move into the other unit. Needless to say that really wasn't the arrangement. Madam WRAAC decided she would spring a visit to check out our quarters. A lot of moving of male items began before she arrived. The visit of the Queen coincided with my marriage and prevented Army guests coming to the wedding as they were required for duties.
Andrea Woods (enlisted 1967)

I had a few boyfriends whilst in the Army, but only two serious. All but one of the relationships were with soldiers. The first serious romance was with a civilian, not a soldier, and I thought he was the greatest thing since sliced bread. That was until I met Colin Boughton. Colin came into my life when I was nineteen years old, just two years almost to the day that I enlisted. I'd seen him in the unit over the previous four months, but, until a small group of us decided to have a night out at the local pub one payday after work, I hadn't really spoken much to him. After all I had this cute boyfriend that I spent every waking minute with, so I wasn't interested in looking further. That is until I spent an evening in Colin's

company and realised that I had been dallying with a young lad who faded into insignificance alongside this worldly and exciting mature man. I knew within a few days that I was hopelessly in love and the feeling was mutual. Colin was fifteen years my senior; a handsome, well-travelled career soldier from London, who had been a Serviceman since he was fifteen years of age, and who had enlisted in the Australian Army the year previously and left England. We booked the church for our wedding two weeks after our first 'date', and were engaged two weeks after that. It took Colin that long to retrieve the money owed to him by his mates, so that he could buy an engagement ring. We were married five months later, on 6 March 1965.
Marie Taylor (enlisted 1962)

I stayed in the Army for six years and would have stayed longer, but I met my husband and in that era you could not stay in after you were married. Later you could stay on, but when I got out it was an automatic discharge if you requested to be married so I got out in 1961. I immediately had a family and in those days you didn't go out to work, you stayed home and looked after your children. The fact that I married an Army officer and therefore was aware of rules, regulations and conditions possibly made my marriage a lot easier. My husband was a Vietnam veteran and he served two tours of duty in Vietnam. He was in the Army for thirty-six years all told. At the moment I am living in the eighteenth house that I've lived in since I was married. I didn't find it a problem but the children now tell us that it was very hard on them. It wasn't so much changing schools as leaving their friends behind.
Esma Summerton (enlisted 1955)

I met him at 31 WRAAC Barracks. He was a National Serviceman in the provost corps. He went to Vietnam in 1968 and he served mainly in Vung Tau. We were married in December 1968. At that time you had to leave the Army because that was Army policy which was unfortunate because I really enjoyed my time in the Army. I loved it, but these things happen. You can't change them.
Desiree Harms (enlisted 1965)

I fell in love in Melbourne on my return (from Vietnam). He was a lovely fellow, we are still friends, but he went off to Vietnam and I still was not ready to settle down. I later married a Canadian. No children by

choice. I stayed in Sydney on discharge and commenced my postgraduate work in renal and intensive care unit (ICU). I had an interesting and varied career. My husband was in the NSW Fire Brigade and he did shiftwork as I did. We seldom saw each other which was good I think. I can't say that Vietnam was in any way responsible, but I was not suited to marriage. I love the freedom of being on my own. I'm just that way.
Ann Hall (enlisted 1967)

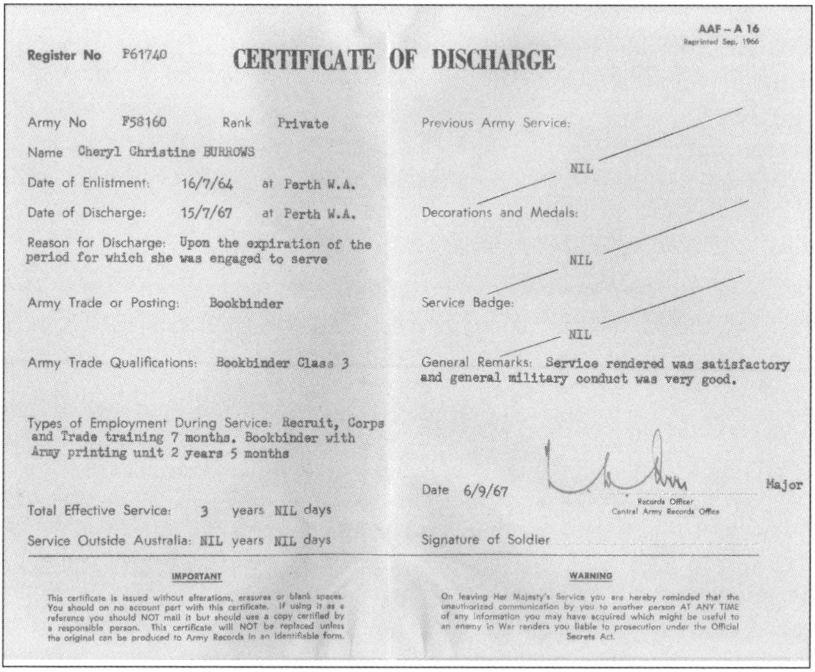

AAF – A 16
Reprinted Sep, 1966

Register No F61740

CERTIFICATE OF DISCHARGE

Army No F58160 Rank Private

Name Cheryl Christine BURROWS

Date of Enlistment: 16/7/64 at Perth W.A.

Date of Discharge: 15/7/67 at Perth W.A.

Reason for Discharge: Upon the expiration of the period for which she was engaged to serve

Army Trade or Posting: Bookbinder

Army Trade Qualifications: Bookbinder Class 3

Types of Employment During Service: Recruit, Corps and Trade training 7 months. Bookbinder with Army printing unit 2 years 5 months

Total Effective Service: 3 years NIL days

Service Outside Australia: NIL years NIL days

Previous Army Service: NIL

Decorations and Medals: NIL

Service Badge: NIL

General Remarks: Service rendered was satisfactory and general military conduct was very good.

Date 6/9/67 Major
Records Officer
Central Army Records Office

Signature of Soldier

IMPORTANT

This certificate is issued without alterations, erasures or blank spaces. You should on no account part with this certificate. If using it as a reference you should NOT mail it but should use a copy certified by a responsible person. This certificate will NOT be replaced unless the original can be produced to Army Records in an identifiable form.

WARNING

On leaving Her Majesty's Service you are hereby reminded that the unauthorized communication by you to another person AT ANY TIME of any information you may have acquired which might be useful to an enemy in War renders you liable to prosecution under the Official Secrets Act.

Figure 46 Certificate of Discharge

My best friend from high school was living in Canberra at the time, and I hung out with her a lot. I actually have her to thank (blame) for introducing me to my future husband. Hubby wasn't in the Services, but he had been in the Police Force, and while he had a good understanding of the discipline required and the shift hours involved, it proved difficult for me to balance family life with work, as I inherited two ready-made children along with him. Since I had only signed on for the minimum

three-year period, I was given the option in 1983 to re-post to Melbourne's 6 Signals Squadron or resign. Romance won out over my career. Three children of our own (and don't forget the grandchildren) and thirty years later we are still together, so I can't say I made a bad decision.
Clare Williams (enlisted 1980)

My Basic Storeman course happened to be on at the time of a major airline strike. I was posted to 1st Training Group, Wacol, Queensland. There were two male soldiers who needed a lift back to their bases and because no aeroplanes were flying I was asked if I could fit them in and take them back, which I did. I am very happy that I did as I ended up marrying one of the boys and we have just celebrated our thirty-sixth wedding anniversary.
Annette Wilson (enlisted 1977)

It was a fairly relaxed lifestyle at Georges Heights, Sydney, with beaches, pubs and the city at our doorstep. Whilst there I met my 'Gunner' John Gray, who was watching me play basketball for the Army at Inter-Service Games. We won, defeating both Navy and RAAF. John and I had a great time together and married in 1958. My position at Victoria Barracks (Vic. Barracks) was made civilian to accommodate me until I left when pregnant with my first child.
Wilma Stralow (enlisted 1955)

I met my husband on our Drivers' Course at Randwick, Sydney, in October 1972. I used to always say I was never getting married. I have girlfriends from school that married and had children not long after they left school which did not appeal to me at all. I think being the eldest of seven must have put me off. Anyway, for someone who was never going to marry, I met Ken in October, we became engaged on the drivers' course and married the following May. That was forty years ago and we are still together. Must say it hasn't always been smooth sailing but we got through. I was posted away to Kapooka on completion of the course. There were five WRAAC on my course and four remained in Sydney and I was posted away. I always felt that they thought our engagement wouldn't last if I was (posted) away. It did make our relationship stronger. I was posted back to Sydney in April as Ken was stationed at Holsworthy and I went to Ingleburn. Back then Ken had to give written consent for me to stay in the Army as did my Commanding Officer (CO). I remained

in the Army for another nineteen months after I was married. I took my discharge as I was five months pregnant and my CO at the time felt it was time to go. When he found out I was pregnant I was office-bound until my discharge. I remember clearly the day of my discharge. It was such an anticlimax. I went to Watsons Bay for the morning and was returned home via Commonwealth car. I was home alone as Ken was away somewhere with the Army. I felt so isolated and I didn't know what I was going to do. We didn't live in married quarters at that time so I didn't know anyone in the area and all my friends were WRAAC. By that stage my life was completely ARMY. I did adjust eventually but it wasn't until after our first son arrived that I finally was able to come to terms with not being a WRAAC. I did miss it for a very long time. I went back to work only three months after my first son was born because I felt so isolated and it was not helping him at all. I have been working ever since in many positions including banking, IT, and for the past ten years as Admin and Financial Controller. I went back to TAFE training in IT.
Fran Sinclair (enlisted 1972)

I was discharged from the Army in August 1966 to get married. My husband's unit was due to go to Vietnam a few months later. However, because his period of service would have been up during the period of time that his unit was in Vietnam he asked if he could sign up for a further twelve months only and he was told no. He was disappointed that he didn't go. On reflection we think that was the biggest mistake. You do what you think is best at the time. I guess my biggest disappointment was that only a few years later women were allowed to remain in the Army when they were married and even if they had children they were still allowed to remain in the Services.
Marion Quick (enlisted 1965)

We'd met these boys when we were at Studley Park Camp in April. Marie's birthday was late May so one of the girls said, 'Why don't we ask those fellows that were at the camp to come to the party?' So they came. We all sat around and talked and some people sang songs. It was pretty primitive sort of entertainment in those days. So that was how I met my husband. After a couple of weeks he asked 'would I like to get married?' I said 'Oh yes, that would be a good idea', so we were going to get engaged. The boys were all getting ready to go to Malaya from 3 Battalion at Ingleburn. He went to Canungra for six weeks' jungle training, then a

couple of weeks after he got back he said, 'If we get married before I go you can come to Malaya'. I thought that was a pretty good idea as well so we got married on 21 September and he sailed on the *New Australia* on 25 September with the battalion. The whole of his unit went away together and some of the wives went on the ship with the men. Other girls went by plane in little groups until I went in January, one of the last ones to go, and we stayed there for two years. We went back to Malaya in 1963 for two years and we came home and that was when the Vietnam War was starting, in the mid-1960s. He went to Vietnam twice with 2 RAR (Royal Australian Regiment).
Margaret Elder (enlisted 1953)

We had a nurse at the barracks who didn't have a family, and when she married all the girls put on the wedding at the barracks for her. I still have a picture of her taken there. Anne joined us at the 60th anniversary in Sydney in 2011. A long time since we had caught up at 30 WRAAC Barracks reunion in 1992.
Betty Hall (enlisted 1963)

My time in the Service was cut short because of the silly ruling that if we wished to marry, women had to be discharged. To make it worse my discharge papers stated that it was at my request!!!!! My time in the Army, though, was very good and I really enjoyed it.
Daphne Gardner (enlisted 1964)

Jenny McLay was working at 1 RTB (Recruit Training Battalion) at Kapooka, NSW when she met Arthur Robertson, a regimental drill instructor.

It was a bit of a fluke that I did meet him because I'd been going out with a chap who I found out was married, so I ditched him and moved on. He kept coming over to the rec (recreation) room and was a bit of a nuisance. This particular night he was more than a nuisance and Arthur wandered over, gave me a kiss on the cheek and said to this fellow, 'She was my girlfriend before you, now bug off'. That's how I met him. I'm thinking to myself, who is this character? You know he wasn't the tallest guy in the world, he wasn't the best-looking guy in the world but he was very easy to talk to. One thing led to another and six weeks later we were engaged. (Later) we had a bit of a break for two months. One thing about the Army social life is that you've got to be careful (because) you can

drink too much. I think it was affecting our relationship there for a while, but we got over that and got married in June of 1966.
Jenny McLay (enlisted 1962)

After march out because our course at Bandiana had been delayed by six weeks we were detached to HQ Training Command and sent to Maritime School, Chowder Bay, Sydney, to live in the officers mess. Because there were only twenty-one staff, breakfast was à la carte. I asked the cook on duty for two poached eggs on a slice of toast, was promptly told 'NO' and got my basic daily ration. I was really annoyed! Two weeks later the cook and I were engaged and have been married for thirty-one years this March.
Catherine Sanders (enlisted 1982)

I was working at Army Headquarters (AHQ), Canberra, when I met John. He was straight off Rookies/Drivers course and as I was a corporal at first he was afraid of me! It didn't last for long, unfortunately. He would drive for the cadets at Duntroon and when they were out on camp, he would bring in the water tanker to be filled and take it back out to them. His routine was that he would stop outside the WRAAC Barracks and I would go out and he would have to help me up into the cab of the tanker as it was an enormous step up! We would chat and 'court' whilst the tanker was filling with water and then he would drop me off at the barracks on his way back out to camp.
Jean Fullerton (enlisted 1963)

I met my husband when he was doing guard duty at Vic. Barracks, Sydney. Since I had to drive him everywhere we spent a lot of time together over a few days (he claims that was the last time I was ever totally respectful and obedient LOL) and we continued to see one another after he returned to his unit. We did manage to rack up a lot of mileage during that time, driving between School of Military Engineering (SME) at Casula, my parents' home, WRAAC School and 101 Transport at Randwick. The mechanic who serviced our cars was really sorry when we got married.
Kim Lake (enlisted 1966)

Wedding Day

Certificate of Marriage

In April 2014, Mike and I will have been married for forty-five years. And they said it wouldn't last! There were times when I thought the critics might have been right but our love won out. We met when we both worked at Vic. Barracks in Brisbane, Mike with Movements and I in the Operations Room. We were both Sergeants and saw each other in the Mess. When we first met he was engaged to someone else. As I'm not good at sharing, I decided that he wasn't for me. However, sometime later the engagement ended, so when Mike asked me out I accepted. Some friends told me I was making a mistake but that just made me want to get to know him even more. Our first date was at the Tower Mill in Brisbane, a revolving restaurant looking over the city, very romantic. Mike is a bit of a romantic. He always remembers our anniversary and my birthday. We were engaged and five months later, we were married on Easter Saturday to give everyone time to get to Melbourne. I borrowed a wedding dress from Nolene Willing, an Army friend, who had been married the year before.
Lorraine Field (enlisted 1964)

When I was discharged from the Army I was about five months pregnant with my daughter. I had married two years previously. It was just at the time when that rule came in allowing married women to stay in the Army. If that rule hadn't come in I don't think I would have got married to be quite honest. I loved the Army and that was my life.
Julie Elverd (enlisted 1960)

My husband was serving in the Armoured Corps located at Puckapunyal. When we decided to get married, unfortunately in those days, a WRAAC had to resign from the forces if she wanted to marry. To lose all of my training and all of my military life because I wanted to get married was a terrible blow.
Helen McLeod (enlisted 1963)

It was at Pucka that I met my future husband and became engaged in September 1965. He was a member of 1st Armoured Regiment and we shared many happy times at dances, the pool and also attending the Cambrai Day Army Ball. We married in March 1966. I was invited by Matron Potter to stay on after my marriage (this was unusual at the time), but my husband had been posted to Holsworthy, NSW, and I did not think it would be a good start to my marriage to be separated. Little

did I know the Vietnam War would do exactly that. My first son was born in December 1966. We spent the first year in married quarters in Holsworthy (NSW) and then in January 1968 he was sent to Vietnam. I found this time very difficult and it was only the kindness of my older sister in allowing us to stay with her that made it bearable. He returned in January 1969 and we moved to Adelaide, which was his hometown. We went on to have three daughters.
Elizabeth Anderson (enlisted 1964)

By early 1966 plans were underway for John and me to marry as he was headed for Vietnam. I had passed my Operator Keyboard Radio (OKR) course with promising prospects of promotion but, with our wedding on the horizon, I had to forgo any thoughts of promotion to plan for my married life. Because of the ruling during this time that females wishing to marry had to be discharged, I left in early 1966 to marry. I was only able to be in the Army for just short of two years and to this day I feel I was denied my right to serve.
Daphne Gardner (enlisted 1964)

In 1974 I met my future husband who was an Army driver at the Royal Military College, Duntroon. My husband's brother was a Vietnam veteran, his brother-in-law a Vietnam veteran, and his nephew a Special Air Service (SAS) officer. I married in November 1975 and set-up house in an Army married quarter in Duntroon. After being married for twenty years, my husband decided that he did not love me anymore. I was never blessed with children. As I was working for a Commonwealth Department at the time of separation, they asked me to transfer to a country town as their Regional Manager. I accepted the transfer and promotion. About eighteen months (later) a friend introduced me to David Barclay, who was a twenty-year Army officer and a Vietnam veteran. When I returned to Adelaide after the Department I was working for was closed, we commenced living together and have been together for seventeen years. The relationship has been a very happy one. However, like with other veterans, there are significant health challenges and difficulties to deal with in close relationships.
Maria Adamczak (enlisted 1973)

While I was at Watsonia I met and married Greg. We had a short courtship and married on 11 February 1967, coming up to the end of my

three-year engagement with the Army. I actually got married whilst we were on leave. In those days I could have been given a dishonourable discharge for not informing the Army and being discharged before marrying. Matron Page, head of the Nursing Corps, called to see me and said that I was breaking the law. However, she said that she didn't want to know anything about it and wished me and Greg a really happy life together. In those days you couldn't be married and be still in the Army.
Ruth Grace (enlisted 1964)

My early departure from the Army after two years of service was because I had the pleasure of meeting a young soldier who worked at 3 BOD and we decided to get married prior to his departure for Vietnam. Looking back I have no idea why that was such a good idea but it seemed a perfectly good one at the time. We were all of nineteen and twenty-one years old. After his Vietnam service we had postings to Papua New Guinea (PNG), Bandiana (Victoria), Hobart and then back to Bandiana. We were married for twelve years but that life ended abruptly when he died in a car accident in 1980. After the death of my soldier husband and the fact that I had a six year old and an eight year old to raise, care for and provide for, I am sure part of my inner sense of I can do it came from the independence and confidence I had learnt as a young serving soldier. Another significant life experience was to train as a Lifeline Counsellor back in the 1980s. Not only did it assist me with my grief process but I learnt some very valuable skills in listening.
Lyn Morris (enlisted 1966)

I was also the photographer for the security passes and this is how I met my husband, who had just been transferred to 6 Signals. We married in May 1973 and were able to get postings to Brisbane – he to 139 Signals Squadron at Enoggera and I went to 4 Sig at Wacol. We arrived in Brisbane in time for the big flood of 1974. During my time there I was 'on loan' to the Regional Engineers at Enoggera and broadened my knowledge and training. I had signed on for another three years before we went to Brisbane and during my time with 4 Sig I was trying to start a family but I was having a lot of trouble staying pregnant – six miscarriages in all and no time off after - just back to work. This incident depicts the male attitude to females at this time. Finally the doctors agreed that I required time off soon after confirmation of a pregnancy. I had had five miscarriages by this stage and when I fell pregnant again I

asked for immediate leave on the Tuesday. The CO said I could have leave after I had completed the wording under a picture that was to be a gift to some mess. This took me until the Thursday and I was to start medical leave on the Monday. On the Friday afternoon getting into the car I started to haemorrhage. I had a curette on the Sunday and was back at work on the Monday. This was the main reason why I left the Army when I did. After the last one my husband, who had been away in the bush on exercise, told me to get out of the system and be out by the time he returned from another stint in the bush. This is why I didn't complete my six years. At first it was hard as I was really given no choice about getting out. Get out or no marriage! Sometimes I think I was a fool and should have done my twenty years like my sister-in-law, but hindsight is 20-20 isn't it? With the arrival of my son and then my daughter I turned my talents to other things. My main goal after leaving the Services was to stay pregnant and after the birth of my son I joined the Army Reserve as a recruiter for 8/13 Victorian Mounted Rifles. This was a complete change from Signals as they were light armour. I stayed with them for three years until my daughter arrived and that took up most of my time.
Hilary Jones (enlisted 1970)

I met my husband when I was doing my Initial Employment Training (IET) training. He was working at one of the Ordnance Units there. I bought a toy for my sister and was walking to the post office to post it. He was looking out the window of where he was working and he made a passing comment to his mate. Something to the effect of, 'Oh, look at that short person over there. Oh, I'll ask her if she'll marry me'. We had our course break-up at the Rabbitohs' Club in Sydney. We walked from there back to the barracks and we stopped off for a pizza on the way with a couple of others. It was a chilly night and he took off his jacket and put it over my shoulders. He left me at our accommodation block and he went on to his block. Two weeks after we met he asked me to marry him. I told my parents and I said, 'What do I do?' Mum said, 'Well, you don't have to accept. You can just be friends', but I loved Pete at that stage. I thought to myself, 'Well if we're meant to be I will get posted somewhere where we can see each other because if I get posted to Sydney or Darwin which is where I'd asked for, we won't see each other anymore and that will be the end of the relationship'. But I got posted to Melbourne and he was posted to Puckapunyal and so we continued our long distance relationship. He would come down to Melbourne on the weekends and

we'd go to my Mum and Dad's place and he would stay in one room and I would stay in the other. Mum and Dad were good like that. I must admit they were absolutely fantastic and became very supportive of me. I think they were happy that I actually found someone, that their daughter was going to get married and there would be no problems. We were married in April of 1978.
Anna Celi (enlisted 1976)

I didn't have any serious romances, just lots of good friends. I met Graham while at Moorebank. He was at Ingleburn. He was a 16th intake Army Apprentice. Just before I was discharged in 1965 he was posted to New Guinea, so our relationship was by phone and letters. We married in 1968. Graham did not go to Vietnam but he was still in the regular Army so I wasn't completely out of it. He served another five years in the regular and then went into Army Reserve (ARES) and served another twenty years. We've been married forty-four years and don't look like parting at this stage!
Julie Ross (enlisted 1967)

After Kapooka I was posted to 30 WRAAC Barracks which is where I was billeted, but I actually worked at the School of Signals, Balcombe, Victoria, where I met the man who was to be my husband. Unfortunately I had to request discharge for the purpose of marriage. That was the regulation at the time. After marriage my husband was posted to Watsonia. He went to Vietnam in May of 1967. Previous to his leaving he had pre-embarkation leave. We went to Queensland to meet his parents – me for the first time, him to say goodbye because he was going off to war, which he duly did, but he left me with his parents. I was twenty years old and in those days I was not shy so much as reticent and I really didn't know these strangers. I found myself pregnant and hormonal and I cried and I cried and I cried. In desperation his poor mother sent me home to my mum. My first son was born while his father was in Vietnam. Benjamin was three months old before his father returned. There was a lot of adjustment to do on both parts I have to say. When Benjamin was ten months old we were posted as a family to Singapore. The two other boys were born after we came back. We were almost three years in Singapore. That was the beginning of the Australian Army's withdrawal from Singapore. While we were there I was to experience a way of life I never would have had the opportunity to in other circumstances. It was

very much run on the British Army way of life for wives. We were referred to by our husband's rank before Mrs, so I was Corporal Mrs Minton and the pecking order was very strict. The CO's afternoons had to be attended, the card parties had to be attended and I found it very stifling and rigid. So in my own quiet way I revolted and opted out. That was a bone of contention as well because I wasn't toeing the party line. But I feel if you're in a foreign country it behoves you to learn how the people who live there live and try to get along with them as best you can which is what I did.
Sheila Cusack (enlisted 1964)

I didn't work in Perth but I did have my first child in 1969 and a second one in 1971, both were born in Perth. I went with Greg back to Melbourne on one of his courses. We rented a house for a couple of months. However, when you're an Army wife I guess that means time managing family by yourself whilst they're away. On one occasion I had a very good friend of mine come to stay, which was wonderful. She was a person that I had shared a room with at WRAAC School, so yes; it's amazing, you meet some terrific friends being in the Army.
Ruth Grace (enlisted 1964)

Conditions of service changed to allow women to remain in the Army after they married. Prior to this you had to resign as soon as you married. Some women were disappointed about this change as they were looking forward to ceasing work once they married. A changeover period of time came in to allow for women to resign upon marriage if they could prove their intention to do so within the next six months. This was only an interim measure and not continued. You still had to resign if you became pregnant and there was no guarantee you and your new spouse could be moved together. Maternity leave came in and it seemed many women became pregnant not long after. There was an expectation you would work until about six weeks before the birth and have a total of twelve months off, using a combination of twelve weeks' maternity leave, annual leave and Leave Without Pay. As soon as you became too big for your uniform you wore civilian clothes.
Christina Morgan (enlisted 1975)

In 1967 I was eight months pregnant with my second son and living in Wagga Wagga, NSW, when my husband got a posting to Port Moresby,

New Guinea. His family wanted me to move to Sydney so they could take care of us while we waited for a married quarter. We were told that would probably only take about three months' waiting time, which in the scheme of things didn't seem too bad. Our older son was nearly seventeen months, so I was about to have two babies to look after. My baby was born a week after my husband left, which was hard, but that was the way things were back then, with so many fathers going off to Vietnam at the same time also. Staying with my in-laws was sometimes a little strained so I did my best to be nice and do everything I could to keep them happy, but it probably wasn't easy for them either.

As time went on, the limited prospect of getting a house anytime soon was causing more strain on the family dynamic. We were told that priority was given to returning Vietnam vets and it was based on a points system. When our son was three months old my husband came down to Sydney for three days before flying back to New Guinea, so at least he got to see his baby briefly. A month later my parents paid for a flight to Perth so we could stay with them until we got our married quarter. This stay turned out to be another four months, but it doesn't matter whether you're staying with family or friends, it's very difficult with two babies. My worst experience, apart from giving birth to my son alone and trying to convince the hospital staff that I was actually married, was the night flight from Sydney to Perth with a two-hour stopover in Adelaide with two screaming babies. I look back now and realise the experience made me very strong and resilient.
Chris Cain (enlisted 1964)

I met my first husband (he was the butcher at 1 Supply), and after our engagement I had to apply for discharge as we weren't allowed to stay in as married women. I was not happy to leave the Army, but as my then fiancé was a reluctant Nasho (National Serviceman), he had no desire to keep in touch with anyone related to his service. The marriage broke down after a few years and if I had realised I was able to, I would have re-enlisted. As things turned out, a few years later, I met my second husband, Noel, and he was still posted to 21 Construction Squadron at Pucka. We lived in Seymour, Victoria, and then moved to the married quarters in Pucka and it seemed like my life was almost repeating itself. He was then posted to Enoggera (Queensland) and I was back in another old posting area again! We even visited an ex-WRAAC who was with me

at 1 Supply and one of the drivers who used to skylark from his truck as he drove by my windows! We're all still in touch.
Kay Owen (enlisted 1964)

As it turned out, I met my future husband shortly after I got to my first posting at Puckapunyal. He was a civilian working in the Pay Office but left to work in Papua New Guinea four months later. Although it became possible to stay in after marriage, I was going to live in Port Moresby after I married in December 1970, so had to take my discharge. As it turned out, I should have stayed in and kept the good times rolling as my marriage failed and I found myself back in Australia with two beautiful babies. Nine years after joining the WRAAC I became a member of the Army Reserve, where I worked almost full-time as a driver and driving instructor at 4 Training Group, Hampstead Barracks (Adelaide). It was great to be part of the Service I had loved so much. Best of all I married one of the instructors from my first course in the Reserve, and we had two children and almost thirty years together before he passed over in 2009.
Deborah Harding (enlisted 1969)

I had a wonderful three years in the Army. I had the opportunity to travel most of Malaysia. An incredible opportunity for one so young. A lot of my school friends worked in chemists and supermarkets and here I was returning from a posting to Singapore. I got out of the Army in April 1972. I had an interview lined up with the Department of Foreign Affairs in Canberra but met up with Tom, whom I had known since Grade 11 at Devonport High School. We had always been friends and often caught up when I was home on leave. We were married on 3 March 1973. Tom and I both loved children so didn't muck about. By the time I was twenty-nine we had three daughters and a son. Sometimes after I'd had the children it irked me when somebody said, 'Oh, you don't work. You're just a mother', when I knew I had run shifts in a communication centre and had served my country overseas during the Vietnam War. I have always been thrilled that I didn't listen to my mother and become a school teacher.
Libby Flemming (enlisted 1969)

This was the write up on our trip in our unit paper. Not bad is it.

16th July 1971.

KL AND BACK

A twelve seater carriage filled with koala bears, Australian bush songs, the latest comics, love stories, lollies, soap bubbles and twelve girls. That was the train ride to Kuala Lumpur to compete in the Malaysian Combined Services Athletics.

Two teams were put together on the Monday, an eight girl tug-of-war and a four girl relay, sent to KL on Thursday to compete Friday. No training but pounds of confidence!

Stars of the tug-of-war were Jay Jay Morrow, Joan Page, Jenny Dixon, Anne Lansdowne, Sally Donaldson, Judy Mohr, Heather Coates, Audrey Inch and Libby Flemming as "The Coach".

The scales were loaded all our way and as the team went up to collect their gold medal, Judy and Heather were still saying, "When do we start."

There was a little more competition in the relay, however Teena Price, Libby Flemming, Zanette Smith and "Flyer" Affleck all ran well for a bronze medal. The gold medal went to the Selangor State runners, reportedly Malaysia's best.

KL Garrison rolled out the red carpet for the girls and the athletics was followed by dinner and a night on the town. There was more than one young lady reluctant to leave KL on Saturday night. All in all, everyone had a ball despite no hot water at the YMCA.

There was one other record broken during the three days. Twelve girls showered, changed and "tizzied" in 25 minutes to go out on Friday night. Keen!!!!!

oooOooo

Chapter 9

Reflections

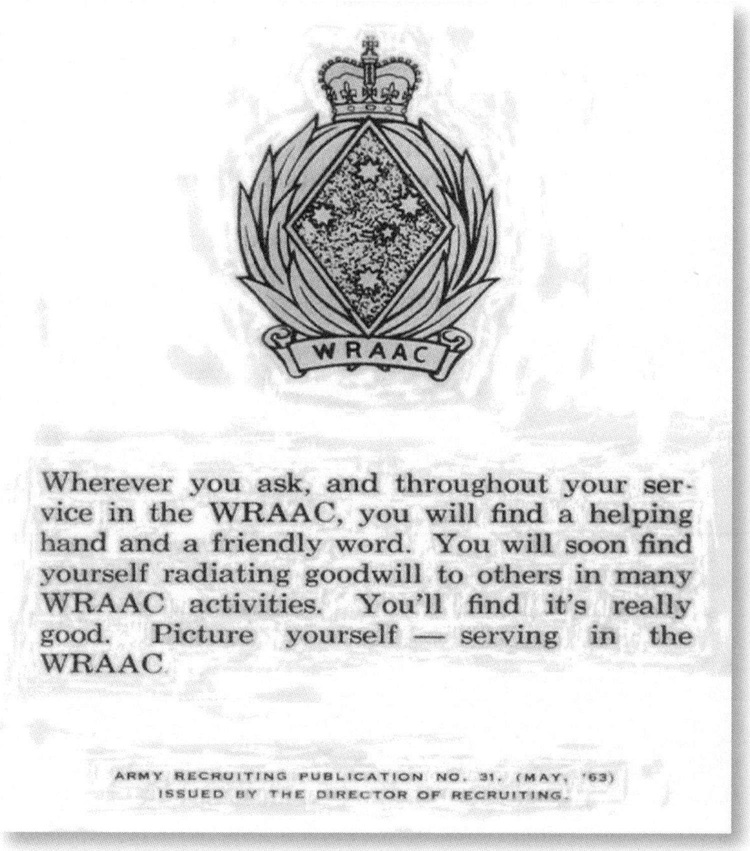

Wherever you ask, and throughout your service in the WRAAC, you will find a helping hand and a friendly word. You will soon find yourself radiating goodwill to others in many WRAAC activities. You'll find it's really good. Picture yourself — serving in the WRAAC.

ARMY RECRUITING PUBLICATION NO. 31. (MAY, '63)
ISSUED BY THE DIRECTOR OF RECRUITING.

The adventurous spirit which led many of these women to join the Army shows through as they tell of their lives after they were discharged. Most of the women interviewed expressed gratitude that they had, after being discharged, been able to get much better jobs because of their military training. The diversity of their later careers shows that these women had the confidence, in some cases, to try something completely new in their post-Army working lives. The training they had received and the responsibilities that they were expected to shoulder as young soldiers stood them in good stead in 'civvy street'. After marriage they put the sometimes short but intense period of Army service out of their minds and concentrated on their husbands and families. Some lost contact with Army friends and so memories of those days were pushed to the backs of their minds.

Coming together again at the WRAAC 60th Anniversary celebrations in Sydney in 2011 revived old friendships and memories, and new acquaintances were made because of shared experiences. The former soldiers were eager to share stories. The years disappeared as they reminisced about their youth and their Army service and talked about what they had done since discharge.
Lyn Kane Dale

There are days when I wonder what my life would have turned out like had I made the choice to stay in the Services. Most of my memories are fond ones, and I would love to go back, knowing what I do now, and being the person I am today, because I'm sure I would make more of the opportunities and challenges that are part and parcel of Army life.
Clare Williams (enlisted 1980)

I felt led to join the RSL in Ulverston (Tas.). I plucked up courage and fronted to one of their meetings, kicked the walking sticks out of the way to get in the door, and their first suggestion was 'Shouldn't you be joining the Women's Auxiliary?' I told them I could cook scones with the best of them but I wasn't the wife of a returned soldier. I WAS the returned soldier. The largely male membership has been very supportive. I trained as a pensioner welfare officer and have been active as volunteer with the Central Coast Veterans' Support Centre, as welfare coordinator for the RSL, and I spend time visiting our veterans in hospital. Then the journey brought me to be elected as sub-branch President - the first woman in our area to be elected to this position in over ninety years.
Libby Flemming (enlisted 1969)

I enlisted in 1965 and finished as a WO2 Instructor upon my discharge in 1970 to marry the 'boy next door'. We left his family dairy farm after a

year or so and eventually started our own company in the timber industry. We have three children, two boys and one girl, and nine grandchildren. My Army career was locked away in a case a few months after my discharge. I had taken my Discharge Certificate into one of the Albury ex-Service clubs and asked to become an ex-Service member. The reply I got was, 'Get your hubbie to join and we will let you in on his Ladies' Card'!!! Needless to say that did not happen and I don't set foot in the place. Only opened my case a couple of years ago when we moved into our retirement home. Even after forty years the memories came flooding back.
Shirley Ball (enlisted 1965)

One weekend in the late 1970s I was working at the Albury (NSW) barracks and a group of us in uniform decided to have lunch at a local ex-Service club. We had ordered our lunch and the guys had shouted the first round of drinks so, when we required another drink I went up to the bar to buy the next shout. As I approached the bar the barman yelled 'Get off my floor!' I said 'Excuse me?' Once again he yelled at me to get off his floor. Stunned, I looked at him with a question on my face and he then yelled that females were not allowed to approach the bar. By this time the guys from my group had joined me and they told the barman to back off. Wasn't this an ex-Service club? Couldn't he see that I was a serving member by my uniform and rank? The guy still refused to serve me and by this time the other females with us decided that they had had enough so we all left, without finishing our meal and with the threat that none of our mates would patronise this club again. I have not been back since.
Hilary Jones (enlisted 1970)

During my time serving in the Army Reserve (after leaving the regular Army) I have always found that to be accepted as a female in a man's world you had to be good at your job and do whatever was asked. I have found that the male soldier was always willing to help the female soldier if she was willing to have a go at any activity. During my time in the Army Reserve I have met lots of people from all walks of life, experienced many varied and different experiences that are not found in everyday life and have travelled to most parts of Australia. I have been fortunate to have served and been given so many opportunities.
Liz Stone (enlisted 1967)

Being in the Army gave me a sense of belonging. I still have that feeling. It was a very positive experience and a massive learning curve, which gave me great organisational and time management skills.
Fran Sinclair (enlisted 1972)

I never expected to have a career. I have continued doing clerical work and changing jobs every few years. Army service broadened my horizons and gave me a wealth of memories and also the ability to go anywhere and feel completely at ease.
Christina Morgan (enlisted 1975)

In my day if women had been able to do the same training as the men I would have wanted to do that. The training with weapons would have interested me a great deal, especially training with some of these new trucks and even tanks, because I'm a bit of a tomboy at heart. I'll have a go at anything.
Desiree Harms (enlisted 1965)

I took my discharge after three years as the powers that be insisted on playing my husband and me against each other, putting us both on the same promotion courses (we would have killed each other) and making empty promises of fast-tracking my military career. Basically I was given the choice - my husband or the Army!!! I may have stopped wearing the uniform but spent the next twenty years going from one unit to the next with Nigel. We both became civilians in December 2000. Now the next generation carries on the tradition.
Catherine Sanders (enlisted 1982)

When I left the Army my husband was posted to Vietnam within a couple of months. I then had to go and find a civilian job. I'd only worked in a potato chip factory or a Four Square store before my military service but the training that I received in the Army stood me in such wonderful stead that when I got back to Perth, while waiting for Walter to return from Vietnam, I got a job as a Settlements Officer with MLC Finance. I had a really well-paid, wonderful job thanks to my training in the Army. It also gave me a lot of self-confidence and an ability to deal with life that I possibly didn't have to such a great extent beforehand.
Helen McLeod (enlisted 1963)

I did six years. In those days we had to come back to our state of enlistment six weeks prior to discharge, so I came back to WA. I lived at home and it was a very, very hard decision for me. I nearly re-enlisted but I decided to get out and I had a few months at home before I went to England to live and work. I ended up with a lovely job I don't think I would have got if I hadn't had that extra confidence and the camaraderie that I got from the Army. You only have to be in the Services to know what that is like. I went to Sydney for the 60th reunion and was with a lot of women that I didn't know – I met a few that I did – but just that being there for people was good. I think if women had done their training with men I would still have joined up. I think that now the opportunities for women in the Defence Force are far, far better. I know there's a lot of controversy going on at the moment about bastardisation but I still think I would have joined up.
Christine Negus (enlisted 1963)

I guess I never really left the Army as my husband continued to serve and after his twenty-one years became a reservist. We still maintain strong ties with our Army friends and go to reunions when possible. I am also a member of the WRAAC Association in Tasmania. We are fortunate enough to have had two sons and are still in the same marriage after more than forty years.
Andrea Woods (enlisted 1967)

My husband left the Army in 1986 and became involved with the Royal Australian Regiment Association and their infantry soldiers. So I just mucked in. In 1996 we became involved with a group that were establishing a memorial to infantrymen killed in war, inside Gallipoli Barracks at Enoggera, Qld, and I started to work with the men who were doing this work. It was making a great big garden of native trees. Over the next couple of years we worked to make this memorial come to fruition. In 1998 the Governor-General dedicated the memorial and we have kept on working ever since. It's been a labour of love for a lot of people. Since the end of the Second World War we've lost almost 700 men from overseas service in Korea, Malaya, Vietnam and now in Afghanistan and Iraq, the Solomons and a couple in Somalia. They're all infantry men remembered at this memorial. There's a big grove of native Australian trees with a plaque beside the tree with the soldier's name on it. The boys all got together and nominated me for an Order of Australia.

It was a really lovely thing for them to do. It wasn't really my Order of Australia. It was recognition of the work that all of them had done to make this memorial because it is a very special place. I love working with the boys. People talk about post-traumatic stress and all the things that are associated with soldiers. I see it happening. Quite often they'll come and maybe talk – I'm the only lady there most of the time – they'll come and talk to me and tell me little things about their lives that they probably would never have told anybody else. I know lots of secrets.
Margaret Elder (enlisted 1953)

After my three years of Army service I went straight back to my old job for another three years until I married. My employment prior to joining was at a business called Farmers & Graziers Pty Ltd in Launceston, which was to do with all things 'on the land'. I did wages for about 100 employees plus dispatching cheques to the farmers on their sale days and other assorted clerical work. I loved it. After I had our three sons I joined the Reserve (1979-1982). That was a completely different Army, a man's Army, so that wasn't quite the same, although I did have fun! I think my Army experience was very positive. I don't know if I had these traits before joining but I'm sure the discipline was great (for anyone), self-esteem gained and the love of ironing which is still my favourite household chore today, much to everyone's disgust. I think everyone should have a taste of Army life.
Julie Ross (enlisted 1967)

I thoroughly enjoyed driving the Mack and whilst I was a civvy driver at Moorebank I managed to get to drive these trucks all over Australia. I did several trips to Adelaide, carting ammunitions and weapons. I also did a trip to Perth carrying missiles for HMAS Stirling, the Navy depot over there. I was in my element and always wanted to drive semi-trailers. I applied with LinFox when they advertised for drivers and I was taken out for a driving test in a truck. The driver trainer showed me the truck I was to drive and it was an International with a long bonnet. Most of their fleet were Volvos and I hadn't driven one of them. So when I hopped into the International I smiled and said, 'Great. This is just like a pair of comfortable slippers'. He had a very puzzled look on his face. I found out the Volvo was just like a big car and the other one had a crash gearbox and was more difficult to drive. He soon found out that I could drive it with no problem. He laughed when I said I was worried when I came in

because I didn't know how to drive a Volvo. He said, 'If you can drive this International you can drive anything'.
Betty Hall (enlisted 1963)

Figure 47 Bethmay Watson instructing at Wacol Army Barracks, 1990.

Any career that spans twenty years has its highs and lows. Being in the Army for so long there are some regrets and the biggest would be spending so much time behind my desk and missing both the opportunities to train in traditional soldiering and time in the mess to make closer friendships. In a way my military career compensated for the husband and children I didn't have in my life. I was forty-three when I met and married Patrick O'Loughlin, a former Armoured Corps soldier and Vietnam veteran. I became involved with the association of past and present members of WRAAC. For a period the reins of control of the Australian WRAAC Association passed to Queensland and I became the National Secretary. It was a great chance to meet ladies who had served their country as their chosen career. What great stories they told. I've found that the esprit de corps is, as ever, strong. The years may have brought changes but these former WRAAC are ladies still.
Bethmay Watson (enlisted 1975)

I got out because we decided to have a family after six years of marriage. I went on maternity leave and my first daughter was born. I really couldn't go back into the Regular Army as neither of us had relatives to assist us to look after the children. There was no childminding available at the time. I went back into the Army Reserve until my second child was born. It was very difficult resigning from the Army. I hated doing it and still miss it. Later I did my High School Certificate, went to university, got First Class Honours, was invited to tutor and give lectures at Monash University then to RMIT as a lecturer. My particular field is in Australian History and Women's History.
Iris Petrass (enlisted 1961)

I missed the Army life and friends for many, many years. I loved so much about the adventure of Army life - the camaraderie. It has been the best life skill for me to put me on a path to great jobs and it was the best time of my life. Such fun, adventure, but the mateship was extraordinary.

In my current job my boss says to me that he can see that I was in the Army, it has given me leadership skills, life skills, a sense of grounding and being able to make solid, quick decisions for the good of others. I am currently an AO8 in the Education Department, earning over $100,000 per year. Not bad for a grade 10 graduate. I honestly believe it is my Army training and experiences that have given me the confidence that I can do anything. I am well-liked. People comment about my work ethic and my ability to get teams working together. I have had a nomadic life with my Dad being in the Army for over twenty-five years and then myself joining. This has enabled me to adapt and make friends quickly. Last year I started a Facebook Reunion site for Opsigs (Operation Signals) at Cabarlah and Darwin. It was an amazing experience to have seventy-five people sign up within forty-eight hours. Eventually word spread and long lost friends came on line and it was incredible. The reunion was so very special. I have people now whom I consider my dearest friends as we have a history of the most amazing 'Army life' experience that you cannot explain to others, particularly as there was no internet, mobiles, etc. when we were young and isolated from family.
Sandy Jarrett (enlisted 1980)

When you tell people that you'd been in the Army I think they look at you in a different light because you've had the discipline and work ethic that comes with Army life.
Desiree Harms (enlisted 1965)

The training and discipline, the self-esteem gained, helped me with many aspects of my life since my discharge. I tackled many jobs and was always thanked for my style of work ethic. I have gone on to do many things since John passed away. One is being a member of the State Emergency Service (SES) and being chosen to become a Peer Supporter. This is being a counsellor for any member who is undergoing some problems themselves. A Peer Supporter has a big responsibility to the member they are helping as confidentiality is the most important aspect of the position. My Army history was of huge interest when I applied.
Daphne Gardner (enlisted 1964)

I married a soldier – an apprentice to be exact. He joined when he was fifteen and I met him when he was nineteen. He was stationed with 6 Sig Regiment, Watsonia, and then 139 Sig Squadron (Enoggera, Qld) – which was the jumping off squadron for Vietnam but we were lucky as he didn't deploy. That is not to say that his Army career hasn't caused health problems as he is just now going through his second bout of chemo after seven years in remission with leukaemia. We believe this is due to the high levels of electromagnetic radiation he came into contact with during his time at the outstations of 6 Sig Regiment. At Diggers Rest and Rockbank (Victoria) they had transmitters, and receivers were in full swing because of Vietnam, and any initial radio contact came through there. The Army was a positive experience for me and definitely improved my self-confidence and self-reliance. Previously I had been a mousy type of person who faded into the background, but now, while still quiet in some situations, I will stand up for myself if I believe I am in the right. I believe it gave me more insight into raising a family, dealing with medical situations and workplace conflicts. My work has entailed office type positions in many industries and when my son asked to join Army cadets I faithfully took him the forty-five kilometres there and waited for him each night. I joined Army Cadets in 1992 as a Cadet Officer and served for fourteen years. My final three years were as Officer Commanding (OC) the Albury-Wodonga unit. At sixty-two I am now forced into retirement but if I had my life to do over again I believe the

only thing I would change is the time I spent in the Service. I'd have been happy to stay in for longer.
Hilary Jones (enlisted 1970)

I worked with my husband in running his building contractor's business for years and then went into dispatch of semi-trailers, medical centre supervisor and finally the Queensland Ambulance Service as an Emergency Medical Dispatcher tasking ambulance units and choppers. We tried to give something back to Vietnam. Went in 2004 with a medical team and then took my own team with a non-government organisation (NGO) in 2005 to teach pre-hospital medicine. Have been back every year since, bar one. I have linked up with the Nhat Hong orphanage in Saigon and Da Lat, and have Sister Van Nga as a lifelong friend now. We are going over in 2014 again. My husband and I know we are a good team but have had our moments. I am blessed with the life I have had.
Joie Norton-Willson Scheinpflug (enlisted 1967)

I have long-term friendships with a number of WRAAC friends and Army couples. In 1999 I was at a conference in Perth and met up with five other former WRAAC friends. We were all in our late 40s and all of the women had married soldiers and their husbands were all Totally or Permanently Incapacitated (TPI). It was quite a confronting experience. Those men were not travelling well and the women were doing a sterling job supporting them.
Jayne Rickard (enlisted 1969)

On discharge, I had no trouble in securing positions. When they were aware I came from an Army background it seemed to give me credibility. I am now once again enjoying reflecting on old days and meeting many new friends who were also in the Army. There is most definitely a sisterhood within us. I have many wonderful memories and am so grateful for that time in my life. My understanding of Army life increased with the years and has helped me even today with being able to help other vets and their families here in Toowoomba. I so enjoyed my Army life and give thanks for all it has brought me and would recommend it (as I do!) to anyone, even as a war widow.
Wilma Stralow (enlisted 1955)

Since retiring, David and I have become 'grey nomads', travelling around Australia. I have been active in volunteering whilst also caring for my mother before she passed away in 2010. During my travels I try to meet other ex-WRAAC, usually over lunch at the local RSL. I have been drawn to re-visit my military experiences since retirement, possibly because my great nephews are just starting their military careers and David's affiliations with his battalion and medical challenges make me feel that the Army is still with us and real. Reflecting back on Army life, I believe it contributed to my passion for natural justice, particularly for women. It defined my stance on bullying in the workplace, sexual harassment and abuse of power which likely led me to a career path in Human Resource Development. The Army gave me the opportunity to 'grow up' and learn how to look after myself. It taught me how to push myself through difficult experiences of harassment and realise that the only way through was to 'toughen up' and not let anyone see my pain. All of these experiences and more have enabled me to provide sound counsel when working in Human Resources and in leadership roles. The Army also enabled me to earn a good income and have access to free health care. My ambition for a career in civilian life resulted in attending university part-time and securing positions that provided promotional opportunities. I became reasonably ambitious after leaving the Army. On a negative note, I believe it contributed to my poor body image. I often felt highly embarrassed about my body, which started in the Army and is still evident. I believe my bad relationship with food became an issue as I was trying to get my body to fit my large breasts. The sexual element of harassment was rife in the 1970s and I am sure I was not alone.
Maria Adamczak (enlisted 1973)

When my husband died in 2011 of health problems related to his war service, I came under the umbrella of Legacy, a wonderful organisation set up in 1923 by returned Servicemen to provide care and support for wives and families of incapacitated or deceased Australian war veterans. A new phase of life opened up for me. Legacy paid for me to attend formal media training, and I joined other Legacy War Widows presenting an ongoing weekly radio programme in Adelaide - https://radio.adelaide.edu.au/program/the-legacy-hour/.

This has proven to be extremely therapeutic, giving me a wonderful opportunity to move outside of my comfort zone, keep my brain active,

meet new people and form new friendships. In 2012 I participated in the documentary film, *Lady Soldiers*, and have been truly blessed with many ex-WRAAC returning to my life.
Helen McLeod (enlisted 1963)

If I hadn't joined the Army I don't know where I would have been. Coming from a very small country town, I was so shy that I used to write notes to the shopkeepers and just put 'one loaf of bread or one pound of mincemeat or sausages', and I'd almost fling the note at the person behind the counter because I didn't want to talk to them. The Army did instil confidence, independence and self-esteem, and particularly after doing some of the promotional courses I felt I could do anything. I went a long way with the Army backing.
Julie Elverd (enlisted 1960)

I loved Army life and really enjoyed being an Army wife. I think I 'grew up' in the Army and it taught me a great deal and was a big part of making me who I am today.
Kay Owen (enlisted 1964)

I will never forget my time in the Army. I joined as a shy, reserved young person and when I left I had a much better idea of who I was.
Elizabeth Anderson (enlisted 1964)

A sad thing happened within my career. Each job that I was doing in finance eventually because civilianised. And if I have any regrets it is the fact that towards the end of my career rather than keeping jobs open for people with rank to stay in those jobs, be they male or female, they were civilianising them. Those people in the corps were limited with postings that they could get and promotion as well. I was in during a very good era but towards the end I was glad to leave. I extended my service with some leave and long service leave at half-pay to adjust to being on a lesser income to see out my thirty-first year. Now I've transferred to the Reserves.
Anna Celi (enlisted 1976)

I now hang around with the Australian Women's Army Service (AWAS) here in Bendigo and I am classed as the 'Bus Driver'. The Bendigo RSL has a twelve-seater mini bus, and I have taken the girls over 1,000

since its arrival. Our ladies range in age from ninety-five down to late seventies. They just love going out and it really doesn't matter where. I am proud to be associated with them.
Annette Wilson (enlisted 1977)

I am now retired from a brilliant career as a social worker. I was involved in support groups for Army wives when my children were little, upgraded my studies and returned to the workforce in 1987. After a couple of years in the State Welfare Department I went to the Child Protection Services at the Women's and Children's Hospital in Adelaide, and eleven years after that went on to manage the Women's Social Work Department for nine years. I am very grateful for the opportunities I received in the WRAAC. I left with a professional qualification that has stood me in great stead. I am inclined to think I would have found my way to university but the Army gave me the entrée at twenty-five years of age.
Jayne Rickard (enlisted 1969)

Apart from having about three to four years off when the children were quite young I remained employed in various clerical positions until aged forty. In 1988 I then made a major decision to retrain myself and study to be a Financial Planner. This has turned out to be the perfect job for me as I realised that (a) I really enjoyed working with people and (b) I am in the very fortunate position to be able to provide financial solutions to a wide range of clientele including quite a few ex-Army personnel as I still live and work in the Albury-Wodonga area. I now own and manage a financial planning practice with eleven employees. I joined the Wodonga Rotary Club, serving for twelve years, one of them as President, and am the proud recipient of a Paul Harris Fellowship[27], quite an honour within Rotary ranks. I have never regretted my short two-year career in the WRAAC as I met many wonderful people. I had to use two words to describe those two years they would be fun and friendship, with the friendships still enduring today.
Lyn Morris (enlisted 1966)

Figure 48 Jayne Rickard at the former WRAAC School site. Georges Heights, Mosman NSW, 2014

The Army was definitely a positive experience for me. I joined as a very young and naive seventeen year old and was quickly educated about things I had previously known nothing about. It was always a wholesome experience. I grew up quickly, gained self-confidence and self-esteem. I suppose my Army training was beneficial to me after I was discharged and later went back into the workforce, as I was a more mature and confident person, and had learnt some people skills and the benefits of discipline. I took my clerical skills with me to the Army, so I can't give the Army credit for that, although I guess the clerical course I attended in the Army would have honed those skills somewhat. I guess I'll never really know if having my Service record listed on my CV ever helped me to get a job, but I somehow think that it might have. I adjusted well to being an Army wife. I was eased into it really, as I stayed in my old job as the Orderly Room Non-Commissioned Officer (NCO) as a civilian after I married, until such time as Colin was commissioned and posted out of that unit. Our future postings were to Signal Regiments and I always knew people there from my Service days and met and befriended other Army wives. The Army is a family that I've always been part of, and I've always felt at home around Service people. The Army watched over me while Col was in Vietnam and I didn't miss a single Mess dinner during his absence, being picked up and safely returned home again.
Marie Taylor (enlisted 1962)

I think that serving in the Army, even for only a couple of years, was the best thing that could have happened to me. It gave me a sense of purpose; it enabled me to find and foster a level of confidence and capability that might never otherwise have surfaced. Married to an engineer for so long and living in so many different countries has thrown up situations and issues that I believe I've dealt with pretty well, and that's a direct result of Army training and the fostering of self-motivation. Yes, it was certainly a positive experience and one that changed my life in so many ways, all good. As an eighteen year old, I don't know how I expected my life would go but even with the bumpy times and some tears it's been enjoyable and I wouldn't change my life, experiences, or my lovely husband. Although I do wish that I could find the waistline of that eighteen year old girl who joined the Australian Army.
Kim Lake (enlisted 1966)

The best times of my life were when I was in the WRAAC and the Army Reserve, being able to see Australia free of charge and be paid to drive trucks which was a passion. I believe I have had a fantastic life.
Betty Hall (enlisted 1963)

I had no idea back in 1964 when I was only seventeen and made the important decision to join the Army what a positive impact that would have on my life. The Army taught me discipline, consideration for others, good house-keeping skills, especially ironing. I can still remember my gigs (giggle dresses) so stiffly starched that they could stand on their own. It would take nearly an hour to iron each one. But the most important thing was pride when in my uniform. We, as WRAAC, had a dignity and responsibility to each other as well as representing Australia. Being a WRAAC had a very positive impact on my life and is a decision I have never regretted.
Cherryl Burrows (enlisted 1964)

From the time I settled down to Army life I enjoyed the whole concept. I believe it helped to develop me into a better woman with skills that even today I can see in my work ethic. I cannot stand being late for appointments. I'm always those five minutes early or even earlier. I had so many rewards. Being posted to marvellous cities and meeting new friends, enjoying the entertainment and sightseeing and lastly, being in the uniform. I loved the 1970s uniform. I was proud of being a female soldier.
Denise Darbey (enlisted 1973)

(After my discharge) I was restless for ages but that is me too I guess. I wanted something different and I got it, and still each day I am on the lookout for new experiences and to put the most into each day. (Being in the Army) for me was a very positive experience. I did something special and I met some fabulous people because of that experience. I learned heaps, had fun, fun, fun and would do it all again.
Ann Hall (enlisted 1967)

There are so many wonderful stories, memories and absolutely priceless friendships from the twenty-two months I spent as a WRAAC. It was the time of my life. What a trip! I loved it all.
Deborah Harding (enlisted 1969)

WRAAC
AT THIS WRAAC SCHOOL SITE (1958-1984)
WE REMEMBER THE MEMBERS OF THE
WOMEN'S ROYAL AUSTRALIAN ARMY CORPS
WHO SERVED WITH PRIDE AND DISTINCTION
1951 - 1985
SOLDIERS OF THE QUEEN

The Lady Soldiers

Andrea Woods at the age of nineteen, enlisted in Hobart in 1967 along with her sister, Marilyn. She served just under five years in Victoria, Canberra and Tasmania in clerical positions. In 1970 she married an Army Apprentice and was posted to Hobart while he served in Vietnam. Andrea was discharged in 1972 and had two sons. She continued living an Army life until her husband's retirement. Her interests include spending time with family, belonging to the WRAAC Association, Legacy, Partners of Veterans Association and RSL Women's Auxiliary. She is a Welfare Officer for her RSL Sub Branch and Treasurer of the East Coast Vietnam Veterans (Tasmania).

Ann (Johanna) Dullaart enlisted in Sydney in 1966 aged seventeen. Her postings were to Puckapunyal and Victoria Barracks Paddington, NSW where she enjoyed competing in Inter-Service swimming. After her discharge, Ann followed her dream and travelled around the world for eighteen months. After returning home, she met her husband, Ross, and they continued to travel to exciting places with work. They have a son, a daughter and two adorable granddaughters. Ann worked in Finance before retirement. Both Ann and Ross continue to enjoy travel and spending time with their grandchildren.

Ann Hall did her nursing training at the Mater Hospital, Brisbane 1962-66, then midwifery in Sydney 1966-67. She was in the RAANC from 1967 to 1971 during which time she served at 1 Australian Field Hospital, Vung Tau, Vietnam. After leaving the Army she completed postgraduate training at Sydney Hospital in Renal and ICU and worked in several

Sydney hospitals before her marriage. Ann and her husband, David, travelled extensively, sailed, and mostly had a great time until David's death in 2007. Ann now has a full life of travel, books, food and friends.

Anna Celi enlisted in Melbourne in 1976, four days before her nineteenth birthday. After initial employment training, where she met her future husband, she worked in the clerical field for three years and then in the financial field in Vic., NSW, Qld, and the ACT. Anna served on active service on Operation Solace in Somalia in 1993 as a Finance Warrant Officer. She transferred to the Army Reserve in 2007 and is currently an active member serving at Puckapunyal. Anna married a serving soldier in 1978 and they had four boys. Anna is still happily married and enjoys her six grandchildren and semi-retirement with her husband.

Annette Wilson enlisted in Melbourne in 1977 at the age of twenty and served for nine years working in Q Stores. She married a soldier in 1978 and they have three children and five grandchildren. After her marriage Annette did full-time home duties and has been a volunteer for twenty years at a local primary school. She also drives the Bendigo and District welfare bus for ex-Service women. Annette took up cross country running aged forty-five, race walking at forty-six, and track and field athletics aged fifty. She recently won a bronze medal in the 3000m Walk and a gold medal in 4 x 100m Relay in the Oceania Masters Games.

Bethmay Watson enlisted in the Citizen Military Forces, in Brisbane in 1975 aged twenty-six. She maintained parallel careers, full-time civilian and part-time military, working as a bank clerk and training others in computer-related areas. Now retired,

Bethmay is committed to local women's groups and other community projects. She is a former National Secretary of the Australian WRAAC Association.

Betty Hall was aged seventeen when she joined the WRAAC in 1963 in Victoria as a driver. She was discharged in 1966 when she married an Army Apprentice. They had two children and divorced after thirty-two years. Twenty-one years after her discharge she joined the Army Reserve. Driving trucks in the Army was her passion and this formed a career in transport, driving semi-trailers for a living in civilian life. She is now retired and living in NSW.

Catherine Sanders joined the Army in 1982 in Sydney aged twenty-one. After recruit training she spent two months at Maritime School, Chowder Bay (NSW) and met her husband on her first morning there. After her clerical course at Bandiana she was posted to HQ Training Command in Darlinghurst and moved with them to Georges Heights (the former WRAAC School). Catherine served for three years but although she stopped wearing the uniform she spent the next twenty years going from one posting to another with her husband, Nigel. They have been married thirty-one years in March 2014 and have three children.

Cherryl Burrows enlisted in 1964 in Perth at the age of seventeen. She served three years working at 1 Base Printing Company and living at 31 WRAAC Barracks in Melbourne before going back to WA to marry the 'boy she'd left behind'. They had two daughters, and lived for a time in Melbourne and New Guinea before settling in Perth where Cherryl resumed her career as a bookbinder. Her interests include quilting and reading and she and her husband, Brian, enjoy travelling and spending time with their family including their four grandchildren.

Chris Cain enlisted in the RAANC in Perth in 1964 aged eighteen and served at 7 Camp Hospital, Kapooka, NSW until her marriage to a soldier in 1965. They had several postings and three children before her husband retired from the Army in 1980. Chris took up pottery in 1974 and completed a Visual Arts Degree in 1990. She and her husband live on a property in the southwest of WA where Chris has a successful business working as a Glass Artist in her Studio & Gallery and has won several awards for her work. Chris and her husband, Col, enjoy travelling and visiting grandkids that live in other parts of Australia.

Christina Morgan enlisted in Adelaide in 1975 at the age of seventeen and went on to serve in pay and clerical positions in NSW, Qld, NT, SA and Vic. for the next twenty-two years before retiring to NSW with her husband Neal. Most postings were around three years duration and in civilian life Christina has continued this pattern of changing clerical jobs. She has worked for a wide range of public and private industries and is currently employed in the travel industry. In her spare time Christina enjoys reading, crocheting, gardening, catching up with friends, going to the gym, and travelling.

Christine Negus enlisted in WA in 1963 at the age of seventeen and served for a total of six years. Her first posting was to Duntroon, ACT, before going on to clerical positions in Ordnance in Meeandah, Brisbane and Moorebank, Sydney. She and her husband, David, have been married for forty-two years and have two sons, both of whom have served in the Royal Australian Navy. Christine has been a full-time mother interspersed withperiods in the workforce while also being involved in many community organisations, one of which has been the role of Treasurer at her local RSL sub-branch for ten years.

Clare Williams enlisted in Sydney in 1980 at the age of eighteen and spent three years serving in 135 Sig Sqn in Canberra before leaving the Service in 1983 to marry. Clare and her family moved to Byron Bay in 1984 where they owned a motel and ran the local taxi service for some years. After a brief stint back in Canberra, the family moved to Brisbane where they still reside today. Clare and her husband have raised five children, and have seven grandchildren. Clare has worked for the Queensland Public Service for the past sixteen years.

Daphne Gardner enlisted in Brisbane in 1964 at the age of seventeen, and after recruit school was sent to Balcombe, Vic. on an Operator Keyboard Radio (OKR) course where she met her husband, John. They were married in 1966 before John was deployed to Vietnam. After his return they had two sons and a daughter and eventually seven grandchildren and one great grandson. Daphne lost John due to a long-term illness after forty-four and a half years of marriage. She is involved in SES also a singing group who perform at many places including retirement homes, and many other activities to help fill that empty space.

Deborah (Debbie) Harding enlisted in Sydney in 1969 aged seventeen and served in several units in Victoria as a driver. She lived in Papua New Guinea for two and a half years after her marriage. After returning to Australia she lived in Darwin. After her divorce in 1976 she moved to Adelaide with her two small children, and in 1979 joined the Army Reserve as a driver and driving instructor. She remarried in 1981 and had two more children. Widowed in 2009, Deborah works as a CSO for Bank SA. She enjoys gardening, reading, knitting, needlework and her family which now includes three grandchildren.

Denise Darbey enlisted in Brisbane in 1973 at the age of eighteen and spent twelve years serving in Electronic Data Processing, Operator Keyboard and Cipher in positions in Canberra, Melbourne, Brisbane and Townsville. After her discharge she married, had a daughter and is currently working in the banking industry. Denise is currently working as head of a group of young ladies training them to be tomorrow's leaders. She also works with homeless people on the street.

Denise Dunn enlisted in the WRAAC in 1976 aged twenty-one, and transferred to the Royal Australian Corps of Signals in 1978. She served twenty-four years in the Australian Army in a variety of command, operational, administration, training and capability development positions including two years on exchange with the British Army. Following her military career, Denise joined the Australian Public Service, serving with the Department of the Prime Minister and Cabinet and with the Department of Defence. Most recently, she was Program Manager, Base Security Improvement Program.

Desiree Harms enlisted in the Citizen Military Forces (CMF) in Perth in 1965 aged seventeen years. She enlisted in the Permanent Military Forces in 1967 and served in clerical positions until her marriage in 1968. Desiree and her husband have two sons and three grandchildren. Desiree spent most of her working life in secretarial roles. She enjoys travelling overseas, reading books, going to the movies and local gym.

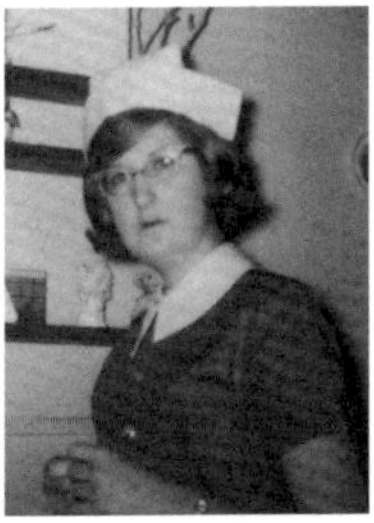

Elizabeth (Liz) Anderson enlisted in the RAANC in Brisbane in 1964 at the age of eighteen. She spent two years nursing at Puckapunyal before discharging to marry. Two years later, her husband deployed to Vietnam. After his return they moved to Adelaide where they had four children before returning to sunny Queensland. Elizabeth made the decision to stay home and raise her children, doing casual work here and there. She now travels the country seeing her six grandchildren, who are all interstate, in-between her commitments to her local bowls club. She and her husband, Mick, celebrate forty-eight years of marriage in 2014.

Esma Summerton enlisted in Adelaide in 1955 aged nineteen and became a drill instructor. She spent six years in the WRAAC in various postings before being discharged in 1961 to get married. She spent the next thirty-six years as an Army wife living with her family in many different areas of Australia. She now lives in South Australia.

Fran Farrell enlisted in the WRAAC in 1963. She was posted to 31 WRAAC Barracks in Melbourne as the barracks' driver and spent five and a half years there doing a variety of duties. In 1968 she was posted to HQ Liverpool area (NSW) and later to 2 Military Hospital (NSW) as Admin and Discharge sergeant. She was discharged in 1971 after serving nine years. Since then Fran has lived and worked on her dairy farm in Yarras, NSW.

Fran Sinclair enlisted in Sydney in 1972 at the age of eighteen and served for two and a half years in NSW in various units as a Stewardess and Driver before taking her discharge to get married. She married an Artillery man and over the next twenty-three years the family followed her husband on many postings until he was discharged and they settled on the Central Coast of NSW. Fran worked in the banking industry for many years before moving into IT. She is now a Finance and Admin Manager and has held this position for the past ten years. She and her husband, Ken, have two sons, three granddaughters and one grandson.

Gloria Coombe enlisted in the RAANC in Adelaide in 1960. After completing recruit training and a nursing course, she was posted to 3 Camp Hospital, Puckapunyal. Five months later she married a soldier and they had four sons. After twelve years of marriage Gloria and her husband divorced and Gloria later remarried and had a daughter. After that marriage ended she reunited with her first husband, Bob. They have nineteen grandchildren and nine great grandchildren. Gloria's hobbies are line dancing and reading and she is now learning acrylic painting.

Helen McLeod enlisted in Perth in 1963 shortly after her seventeenth birthday. She spent three years working in clerical positions in the WRAAC before marrying in 1966 and relocating to South Australia where she still resides. Helen worked in administrative and executive positions in civilian life before retiring in 2001. In 2012 at the age of sixty-six she underwent Media training and joined other Legacy widows presenting a regular weekly radio program called *The Legacy Hour* on Radio Adelaide.

Hilary Jones enlisted in Melbourne in 1970 at the age of nineteen. Her postings included Survey and Signals in Vic. and Qld. In 1973 she married a Signals Cypher Technician and they were posted to Brisbane. After returning to Victoria in 1975 Hilary joined the local Army Reserve unit. She later took various clerical positions and in 1992 became an Officer of Cadets with the Albury-Wodonga unit. She and her husband have been married for forty years and have a son, a daughter and three grandchildren. Hilary left cadets in 2004 to indulge her love of quilting and to help care for her grandson.

Iris Petrass enlisted in the CMF in 1961 aged twenty-six. She served for six years as a Recruit Instructor and Signals Officer before transferring to the Regular Army in 1967. Her postings were as Staff Officer HQ Eastern Command, 4 Military District and 3 Military District, and as Officer Commanding 31 WRAAC Barracks. She married an Army Engineer in 1970. Following discharge from the Australian Regular Army (ARA) she rejoined the Army Reserve until 1980. She later became an academic, retiring in 2009. She is currently patron of the WRAAC Association of Victoria.

Jayne Rickard enlisted in Victoria at eighteen in May 1969. Postings were to Army Recruiting in Perth and Melbourne. In 1974, after marrying an Army Officer, she worked in Army Community Services in Adelaide. In 1975 she was selected to attend UniSA to study social work. Discharged as a sergeant in 1978, she has three children, was divorced in 1992 and remarried in 2006. She has had a lengthy and rewarding career in social work until retirement in 2010. Jayne and her husband, Locky, enjoy sailing, skiing, international travel and golf.

Jean Fullerton enlisted in Sydney in March 1963 at the age of eighteen and served three years in clerical positions in NSW, Vic. and ACT. Upon her discharge in 1966, she became engaged to a soldier who was then posted to Vietnam. They married upon his return fourteen months later and went on to have four children and had been married for forty-four years at the time of his death. Jean spent many years involved in childminding. Her interests are knitting, sewing and grandchildren, and she is involved in two War Widows groups.

Jenny McLay enlisted in NSW in 1962 aged eighteen and served in clerical positions for almost four years until her marriage. After her husband Arthur's death in Vietnam in 1967 Jenny brought up her daughter alone. She never remarried. Jenny worked as a Bookkeeper and Finance Manager until her retirement. Jenny's interests include genealogy, gardening, golfing (past tense), travelling and outings with friends, lots of time with family and walking her rascal dog.

Joan-Marie (Joie) Norton-Willson Scheinpflug enlisted in Brisbane in 1967 at the age of eighteen and served for seven years in Signals in Qld and Vic. She was the first married WRAAC in Qld in 1969. She and her husband, a Vietnam veteran, had three daughters. Joie spent ten years working with her husband in their building business, eight years as a despatch and pallet controller, and a further ten years with Queensland Ambulance as an Emergency Medical Despatcher before retiring in 2009. Her interests include various projects in Vietnam, travelling, reading and spending time with her six grandchildren.

Julie Elverd enlisted in 1960 in NSW and worked in Signals. She married in 1970 and served until 1972 when her daughter was born. After her discharge Julie worked in the fashion industry for ten years and in the public housing sector for twenty years. She retired in 2010. Sadly, Julie's husband passed away in June 2011 after forty-one years of marriage. Julie has been a keen rower for ten years, loves tai chi and yoga and catching up with friends.

Julie Ross enlisted in the WRAAC in Hobart in 1962. Her postings were to 3 Military Police and 21 Cadet Bn, Vic. and in NSW where she worked as a clerk at 2 BOD until her discharge in 1965. In 1968 Julie married Alexander Graham Bourne and they have three sons and six grandchildren. In 1978 she joined the Army Reserve in Launceston. Her greatest achievement there was winning the Best Female Rifle Shot using an SLR and an M60 machine gun. Julie's civilian employment has been mostly clerical though for the past fourteen years she has enjoyed being a casual driver with AVIS Car Rental.

Kay Owen enlisted in Sydney in 1964, aged eighteen, and served in clerical positions in Vic. and Qld until her discharge in August 1968 to be married. After a divorce and then marriage to a serving soldier, Kay and her husband, Noel, raised two sons and a daughter. The family moved to Qld, eventually managing a large hydroponics farm until Noel's retirement. Kay and Noel have eleven grandchildren. Kay's interests include reading, music, knitting and volunteer work with her church, the Red Cross and the Ex-Servicewomen's Association in Bundaberg. She enjoys keeping up with her family and friends and travelling.

Kim Lake enlisted in Sydney in 1966 at the age of eighteen. She served in clerical positions in NSW and Victoria before becoming a driver at 101 Transport at Randwick, NSW. Kim married in 1968, shortly before her husband, Jim, went to Vietnam. On his return they moved to Qld and civilian life. They have lived in numerous countries (the USA, UK, Thailand, Vietnam and, most recently, four years in the Middle East). Kim & Jim have two children and three grandchildren. They live on a small property outside Maleny in the Sunshine Coast Hinterland and have just celebrated their 46th wedding anniversary.

Libby Flemming aged seventeen enlisted in Hobart in 1969. During her three years in the Army she served for a time in Signals in Singapore during the Vietnam War. She and her husband Tom have been married for over forty years. They have four children and nine grandchildren. Libby is currently a Justice of the Peace, Honorary Bench Justice and is past President of the Ulverstone RSL Sub Branch in Tasmania. In her spare time 'Spins' writes children's stories.

Liz Stone enlisted in the CMF in 1967 and in 1970 changed over to the Regular Army. She was discharged in 1971 to marry a Vietnam Veteran. When her husband was posted to Sydney in 1975, Liz was able to rejoin the Army Reserve and has served since then. She divorced in 1987 and has been happily single since then. Liz worked for the Public Service in Defence until 2004 when she took early retirement, and she then worked in Aged Care for a few years. Her interests are travelling, showing dogs, antiques and spending time with family and friends.

Lorraine Field enlisted in Melbourne in 1964 aged eighteen and served in the WRAAC for five years in clerical positions in Vic., WA and Qld before being discharged to marry a soldier in 1969, a year before he went to Vietnam. They had many postings on the east coast and two years in Singapore. Lorraine and Mike have a son, a daughter and six grandchildren. Lorraine worked in many clerical positions as the family moved thirteen times in twenty years. She has always loved sport and her garden and spending time with family. She enjoys being active in local groups especially her church and the local U3A.

Lyn Kane enlisted in Perth in 1964 aged eighteen. She served for five years in clerical positions in NSW, Vic. and WA before marrying a Vietnam veteran and spending the next twelve years living in mining towns in WA. After her marriage ended in 1981 Lyn and her daughters moved to Perth. She worked for twenty-one years in administration at Murdoch University before her retirement in 2009. Her interests are having fun with her grandchildren, spending time with extended family and friends, quilting, reading, travelling and telling women's stories. She has spent eleven months working on this book.

Lyn Morris enlisted in Melbourne in 1966 aged seventeen. She served for two years as a driver in Melbourne and Sydney then in 1968 married a Regular Army soldier who was deployed to Vietnam later that year. Further postings after Vietnam were to Port Moresby, Hobart and lastly Wodonga. Widowed in 1980 Lyn has two daughters. She had a varied administration career for many years, then found her niche in Financial Planning and now owns her own practice in Wodonga. Her interests are gardening, travelling, walking, the odd bit of golf and spending time with family and friends.

Margaret Elder joined the CMF in Sydney in 1953 at the age of nineteen. She married a soldier in 1957 and spent the next thirty years as an Army wife. In 2007 Margaret was awarded an Order of Australia in recognition of the past fifteen years that she has been involved in establishing and maintaining the Royal Australian Regiment National Memorial Walk at Gallipoli Barracks Enoggera, Queensland. Margaret has written a number of poems about Army life and the men who served in the Royal Australian Regiment.

Maria Adamczak was born to post-war immigrants who arrived in Australia through the displaced persons legislation. She enlisted in Adelaide in 1973 just before her eighteenth birthday and spent eight years serving in clerical positions in Canberra, Sydney and Adelaide. Since leaving the Army, Maria has worked in the finance, community and Government sectors in management positions whilst studying at university. She lives with David Barclay, a retired infantry Major. Maria is now retired, volunteers with various organisations, is currently studying writing biographies and enjoys travelling around Australia.

Marie Taylor enlisted in Sydney in September 1962 at the age of seventeen, and served until her marriage in March 1965 to a career soldier. Marie and Col had a daughter and a son and have a young granddaughter. Upon Col's retirement from the Army they relocated to the Sunshine Coast of Queensland, where they still reside. Marie worked for Noosa Council for twenty years, before retiring when diagnosed with breast cancer. She now enjoys good health and her interests include cards, mahjong and gardening. She keeps in touch with ex-WRAAC via the WRAAC Reunited website.

Marion Quick enlisted in the Royal Australian Army Nursing Corps in Adelaide in 1965 at the age of eighteen and served until her marriage. She and her husband, Ray, had four children, one since deceased, and have ten grandchildren and four great grandsons. Marion attended university as a mature age student, gained a nursing degree and worked in this field for a number of years. She and Ray are now retired and living in Adelaide. They spend their time travelling, gardening and just enjoying life.

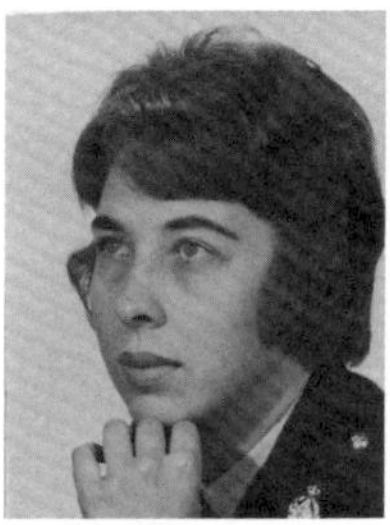

Merle Viles enlisted in Sydney in 1962 aged eighteen and spent seven and a half years at Moorebank NSW, as a Storewoman Tech and Clerk before her marriage to a Vietnam veteran in 1970. She and her husband, Brian, had two daughters and they spent most of their married lives in the Wodonga area. Merle worked for the Australian Tax Office until her retirement in 1998. Since then she has been busy with women's groups, knitting blankets for kids at a women's refuge, reading, cross stitching and learning to quilt. She was widowed in 2008.

Patricia (Trish) Courtney enlisted in Brisbane in 1962 at the age of eighteen and spent twenty-three months in the WRAAC. She served at Mount Martha, Vic., before doing a driving course and was then posted to Watsonia before marrying a soldier. They had four children, one before Trish's husband went to Vietnam, then two more and a stillborn after his return. The family were then posted to South Australia and also had postings to Bandiana, Vic., and NSW. Trish divorced after twenty-three years of marriage and is now very happily married to her second husband.

Ruth Grace joined the Royal Australian Army Nursing Corps in Queensland in 1964 at the age of eighteen and served for three years before her marriage in 1967. She and her husband, Greg, have a son, a daughter and three grandchildren. They moved to Canberra in 1978 and Ruth joined the Public Service. Ruth and Greg eventually resigned from their jobs to develop their farm and cattle stud. They are now retired and living at Coffs Harbour.

Sandy Jarrett enlisted in Newcastle, NSW, in May 1980 at the age of seventeen and spent six years as an Operator Signalwoman in RASIGS. Her postings were to Cabarlah, Qld and Darwin. Sandy has three sons and is currently a Business Services Manager at a large North Queensland High School. Sandy recently was on the organising committee to facilitate a reunion for past serving members at Cabarlah, near Toowoomba. The Reunion Facebook page was instrumental in connecting 300+ serving members, from the three Services and resulted in an amazing sharing of old and new stories.

Sheila Cusack enlisted in Adelaide in 1964 at the age of seventeen and requested her discharge to marry after two years. She and her husband had three boys and were divorced after eleven years. Sheila later remarried and had a daughter. Her three sons all enlisted in the Army. Sheila and her husband, Roger, now live in Port McDonnell in SA and operate a Bed & Breakfast. Sheila qualified as a Chef after fifteen years in TV administration. She is an avid reader, loves to travel and spend time with family. Her interest in photography and painting is ongoing.

Shirley Ball enlisted in Melbourne in 1965 aged nineteen. She served for five years doing Topographical Drafting and Model Making, admin at 2RTB & Army Recruiting and as an instructor at WRAAC School before being discharged to get married in 1970. She was an Audio Visual Technician for eleven years and then worked with her husband in their company in the timber industry from 1989. Their three children have filled their lives with love, laughter, basketball, netball, marathons and music. Shirley is now busier than ever with grandchildren, gardening, knitting and travel.

Trish Webb enlisted in Melbourne in February 1968 at the age of eighteen. She served as a GD at WRAAC School and then in Signals at 5 Sig. Regt. Dundas (Sydney). Trish took her discharge to marry in 1969 as her fiancé was being posted to South Vietnam. She divorced in 1982 and remarried in 1986. Trish has five children and three grandchildren. She spent most of her working life in various positions including Teachers Aide, Tutoring, Correctional Services and Detention Services. Trish is currently a Civil Celebrant and JP. She enjoys scrap-booking, keeping up with friends and travel.

Wilma Stralow enlisted in NSW in 1955 at the age of eighteen and her recruit training was at WRAAC School, Queenscliff, Vic. before being posted back to Sydney and working as a clerk at 2 Base Ordnance Depot. She was later posted to Victoria Barracks, Sydney, as a Draughtswoman. In 1958 Wilma married and her husband John later served in Vietnam. They had two children and eventually five grandchildren. John died in 2003. Wilma is involved in helping other veterans and their families in Toowomba, Qld.

Acronyms

AAP	Australian Associated Press
ADF	Australian Defence Force
AHQ	Army Headquarters
AIF	Australian Imperial Forces
APO	Army Post Office
ARA	Australian Regular Army
ARES	Army Reserve
ASCO	Australian Services Canteens Organisation
AWAS	Australian Women's Army Service
AWOL	Absent without leave
BOD	Base Ordnance Depot
CARO	Central Army Records Office
CB	Confined to Barracks
CMF	Citizen Military Forces
CO	Commanding Officer
COMMCEN	Communications Centre
DORA	Dental Operating Room Assistant
DVA	Department of Veterans' Affairs
EN	Enrolled Nurses
GD	General Duties
HSC	High School Certificate
IET	Initial Employment Training
MP	Military Police
NAAFI	British Navy, Army & Air Force Institutes
NCO	Non-Commissioned Officer
NOK	Next of Kin
NOTICAS	Notification of Casualty
OC	Officer Commanding
OCS	Officer Cadet School
OKR	Operator Keyboard Radio
OR	Other Ranks
PT	Physical Training
PTSD	Post Traumatic Stress Disorder
Q Store	Quartermaster's store

R&Q	Rationed and Quartered
R&R	Rest and Recreation
RAAF	Royal Australian Air Force
RAANC	Royal Australian Army Nursing Corps
RAE	Royal Australian Engineers
RAEME	Royal Australian Electrical & Mechanical Engineer
RAF	Royal Air Force
RAN	Royal Australian Navy
RAP	Regimental Aid Post
RAR	Royal Australian Regiment
RDI	Regimental Drill Instructor
RMH	Royal Melbourne Hospital
RN	Registered Nurses
RSM	Regimental Sergeant Major
RTB	Recruit Training Battalion
RTO	Regimental Travel Offices
SAS	Special Air Service
SG1	Selection Group Rating 1
SLR	self-loading rifle
SME	School of Military Engineering
SMG	submachine gun
SSM	School Sergeant Major
TPI	Totally & Permanently Incapacitated
WAAAF	Women's Auxiliary Australian Air Force
WO1	Warrant Officer Class 1
WRAAC	Women's Royal Australian Army Corps
WRANS	Women's Royal Australian Naval Service
WWG	War Widows Guild
YWCA	Young Women's Christian Association

Notes

Chapter 2

1. Passing Out Parade, sometimes also called the March Out Parade, is an occasion to mark the completion of the recruit training course and is a ceremonial parade which family and friends are invited to attend.

2. Being given the command 'Right dress', the front rank (of three ranks) of soldiers takes one step forward, turns their eyes to the right, makes their left hand into a fist to the left, then extends the arm to the left while shuffling their feet sideways so each soldier is just touching the shoulder of the person to their left. This enables all ranks to be a standard distance from each other.

3. So that the women could perfect their saluting skills they had to salute the flagpole every time they passed it.

4. To square gait is to march or walk with the arm swinging forward at the same time as the corresponding leg i.e. right hand, right leg.

5. Beds had to be barracked every week, when the bed linen was changed. The blankets had to be folded in a certain way and placed one on top of the other with the pillow on top of them before having another blanket wrapped around them so that the 'parcel' looked rather like a sandwich of different layers. On top of this was placed the sheet and pillow case which were to be sent to the laundry. Each week a packet of sanitary napkins was placed on each bed when the General Duties (GDs) came to change the linen. By the 1960s many girls used tampons but the napkins were not wasted and came in very handy for applying polish to the floors.

6. The buttons on the twill winter uniform jacket, greatcoat and beret were on a shank which was pushed through a type of buttonhole on the garment and fastened on the other side by a safety pin through the shank. This allowed the buttons to be removed easily and quickly for dry cleaning and ironing the garment.

7. *See note 1.*

8. A bivouac is a temporary camp, especially a military one, made out in the open with little or no equipment.

Chapter 2b

9. Lisle stockings were worn in the early 1950s.

Chapter 3

10. A supernumerary is associated with a regular body or staff as an assistant or substitute in case of necessity

Chapter 4

11. A donga is a portable prefabricated structure used as a temporary office on a building site, sleeping quarters at a mine, etc.

Chapter 5

12. When National Service was introduced in Australia in the mid-1960s small balls with dates on them were drawn from a barrel and any young man whose birthday was on one of those dates was conscripted into the Army.

Chapter 6

13. Julie was the highest ranking soldier on duty at the communications centre that night.

14. Kangaroo exercises involve training in remote areas in preparation for serving overseas.

15. Regimental Travel Offices (RTO) were situated on railway stations in all the capital cities in the 1950s and 1960s.

16. A padre is the military colloquial term for a chaplain.

17. To form up in the Military means to draw up in lines or in formation.

Chapter 7

18. At one time during the Vietnam War the postal service union refused to handle Servicemen's mail and 'wharfies' refused to load supply vessels for the troops overseas.
19. Source: 'New twist on Rowe draft claim'

    Alan Howe *Herald Sun* May 22, 2008
    http://www.abc.net.au/mediawatch/transcripts/0816_rowe2205.pdf
20. The departure from Vietnam of Greg's unit had been delayed and he arrived home four or five weeks later.

21. Before the advent of the 'Pill' in the late 1960s single women did not have access to reliable contraceptives and, in many cases, were ignorant about what they could do to avoid pregnancy. At that time, for an unmarried girl to get pregnant was regarded as a disgrace and many young women were abandoned by their boyfriends and/or disowned by their parents when they told them of their predicament. Abortion was illegal though some resorted to this method of dealing with an unwanted pregnancy but for most young unmarried women their only option was to go into a church-run home for unmarried mothers and then give up their baby for adoption. There was no government support for women who wanted to keep their babies and a great deal of stigma attached to women producing an illegitimate child. During my time in the Army I knew many girls who were devastated to discover they were pregnant. They would be discharged from the Army and, if their families did not support them, they faced a very uncertain future and the trauma of giving up their baby.

    One girl I knew, whose soldier boyfriend reluctantly offered to marry her after she discovered she was pregnant, told me that during an argument a few months after their marriage he said to her 'What more do you want from me? I did the right thing and married you, didn't I?'

22. A 'Housewife' kit was issued to male serving members so they could do running repairs such as replacing buttons and sewing on stripes.

23. *med·e·vaced, med·e·vac·ing, med·e·vacs.* Air transport of persons to a

place where they can receive medical or surgical care; medical evacuation.

Chapter 8

24. My former husband told me that on several occasions after the aerial sprayings his platoon (and others) had gone on patrol into the jungle while the leaves on the trees were still wet from the defoliant. The leaves then fell onto the ground and onto our troops who slept on groundsheets and walked through the jungle with the defoliant saturating their clothes. My husband was later medecaved from Vietnam several months short of his tour of duty as he developed what the Army doctors called 'tropical ulcers' on the skin of his back, shoulders and neck. The scars from these 'ulcers' looked like burns tissue and it took many years for Repat doctors to acknowledge that his skin condition had been a result of his service in Vietnam. The symptoms from exposure to Agent Orange have had long term effects on the health of a large number of these men and some have died at a relatively young age, with their wives subsequently being declared war widows. Statistics show that a large number of children of Australian and American veterans of the Vietnam War have had birth defects and ongoing health problems because of their fathers' exposure to Agent Orange.

25. The former 31 WRAAC Barracks in St Kilda Road in Melbourne is now a boutique hotel The Hotel Charsfield.

26. A tour of duty in Vietnam was for 12 months.

Chapter 9

27. Paul Harris Fellowship - usually this is given to recognise an outstanding commitment to the Club or the community.

Bibliography

Bomford, Janette M. 2001. *Soldiers of the Queen: Women in the Australian Army*. Melbourne: Oxford University Press.

Biedermann, Narelle. 2004. T*ears on My Pillow: Australian Nurses in Vietnam*. Milsons Point, NSW: Random House Australia.

Defence Widows Support Group – ACT. 2012. *Here, There and Away: Voices of the Families of Those Who Served*. Newport, NSW: Big Sky Publishing.

Goldsmith, Betty, and Beryl Sandford. 1990. *The Girls They Left Behind: Life in Australia During World War II – The Women Remember*. Ringwood, Vic.: Penguin.

Bassett, Jan, ed. 1998. A*s We Wave You Goodbye: Australian Women and War*. Melbourne: Oxford University Press Australia.

Lamensdorf, Jean Debelle. 2006. *Write Home for Me: A Red Cross Woman in Vietnam*. Milsons Point, NSW: Random House Australia.

McHugh, Siobhan. 1993. *Minefields and Miniskirts: Australian Women and the Vietnam War*. Sydney: Doubleday.

Scott, Jean. 1986. *Girls with Grit: Memories of the Australian Women's Land Army*. Sydney: Allen & Unwin.

Winter, Peter. 2003. *The Year I Said Goodbye* Kent Town, SA: Wakefield Press.

Your notes